Presenting Service:

The Ultimate Guide for the Foodservice Professional

Lendal H. Kotschevar, Ph.D., FMP

Valentino Luciani, CHE

National Restaurant Association
THE EDUCATIONAL FOUNDATION

Disclaimer

The information presented in this book has been compiled from sources and documents believed to be reliable and represents the best professional judgment of The Educational Foundation. However, the accuracy of the information presented is not guaranteed, nor is any responsibility assumed or implied by The Educational Foundation of the National Restaurant Association for any damage or loss resulting from inaccuracies or omissions.

Marianne Gajewski, Director, Product Development

Lisa Parker Gates, Manager, Curriculum Product Development

Virginia Christopher, Manager, Product Development/Production

Kate Sislin, Project Editor, Academic Products

Jennifer Hulting, Project Editor, School-to-Work

Elizabeth Witzke, Assistant Editor

Deborah Cayce, Production Editor; Interior Design

Patty Smith, Art Director

Hafeman Design Group Inc.; Cover Design

All photographs in this book courtesy of Stephen Graham Photography and Aran Kessler Photography, except where otherwise noted.

Sidebars in this book are from the following reference books, all published by The Educational Foundation of the National Restaurant Association.

Bar Code®: Serving Alcohol Responsibly, Manager Resource Book. ©1996.

The Buck Starts Here: Front-of-the-House Profitability, Participant Notebook. ©1996.

Customer Service: Manager Handbook. ©1992.

Aware™: Employee and Customer Safety, Manager Coursebook. ©1996.

Foodservice Security, Manager Handbook. ©1993.

SERVSAFE®: Serving Safe Food Certification Coursebook. ©1995.

Library of Congress Catalog Card Number: 96-78408

Inventory Code: MICT413

ISBN Number: 1-883904-67-6

10 9 8 7 6 5 4 3 2 1

Acknowledgements

The production of this book would not have been possible without the expertise of our many advisors and manuscript reviewers. The Educational Foundation is pleased to thank the following people for the time and effort they spent on this project.

Adam Carmer, University of Nevada, Las Vegas

Celia Curry, New York Restaurant School

Ed Debevic's, Deerfield

G. Michael Harris, Bethune-Cookman College

Mike Jung, Hennipen Technical College

Jeanette Kellum, Romano Brothers Beverage Company

Peter Kilgore, The National Restaurant Association

Lettuce Entertain You Enterprises, Inc.

Vickie Parker, Brinker International, Chili's

Lawrence D. Posen, FMP, Eurest Dining Services

Arthur Riegal, Sullivan County Community College

Vincent Rossetti, Nordstrom's, Oak Brook

Peter Simoncelli, Four Seasons Hotel, Chicago

Diane Sinkinson, Cape Fear Community College

Will Thorton, St. Philip's College

Mike Zema, Elgin Community College

Table of Contents

An Introduction to this Text

This book, and its course components, are the latest additions to the ProMgmt. program. ProMgmt. is a college-level curriculum driven by industry research and academic experience, and created by The Educational Foundation of the National Restaurant Association.

This book is accompanied by a student workbook, containing exercises based on each chapter. There is also a study outline and an 80-question practice test in the workbook, which prepares students for the final exam. Students who successfully complete the course will earn a ProMgmt. program certificate.

The eight chapters in this book cover the basics, as well as advanced topics that future foodservice managers need to know, to give successful table and customer service. In addition, three modules, which are not required parts of the course and not on the final exam, cover bar and beverage service, table etiquette, and classic service styles. An instructor may choose to assign one or more of these modules to enhance the course.

We are confident that the information in this course will help build essential skills for a successful foodservice career. Excellent customer service is crucial to the success of any operation, yet it is frequently overlooked by managers and employees. This book is a tool to help build better service practices throughout the foodservice industry.

Preface

Every author, in writing a book, has a purpose. We had two: (1) to detail what managers and servers should know and do to serve foods and beverages competently, and (2) to indicate how one can make a professional career out of serving others. Too frequently the act of service is considered a menial job, one that people do only to make money until a better and more respected job comes along. On the contrary, the foodservice industry enjoys among its millions of employees, many experienced, talented, professional servers who contribute much to the industry's overall success.

Serving people is difficult and demanding work, but the rewards outweigh the challenges. Education, training, and a professional attitude are the ingredients needed to harvest those rewards. With this comprehensive textbook, we hope to educate future servers, supervisors, and managers in the techniques and demeanor of professional service.

Just as a master craftsman takes a raw, crude stone and turns it into a beautiful, sparkling gem, so does an editor take a manuscript and turn it into a published book. The authors would like to thank the editors at The Educational Foundation for their assistance in the development of this book.

Lendal H. Kotschevar, Ph.D., FMP
Professor Emeritus
School of Hospitality Management
Florida International University
North Miami, Florida

Valentino Luciani, CHE
Lecturer/Foodservice Management
College of Hotel Administration
University of Nevada, Las Vegas
Las Vegas, Nevada

About the Authors

Dr. Lendal H. Kotschevar, Ph.D., FMP, has authored or co-authored more than a dozen books. He was one of the pioneers responsible for establishing the discipline of hospitality management. He holds a Ph.D. from Columbia University, and has taught at the University of Montana, Michigan State University, the University of Hawaii, the University of Nevada at Las Vegas, Haifa University in Israel, the Centre International de Glion in Switzerland, and most recently, Florida International University. Dr. Kotschevar, who has lectured widely, received the Meek Award from the Council on Hotel, Restaurant, and Institutional Education (CHRIE) and is also a diplomate of The Educational Foundation of the National Restaurant Association.

Valentino Luciani, CHE, is a lecturer in Foodservice Management in the William F. Harrah College of Hotel Administration, University of Nevada, Las Vegas. He also teaches at the Ecole Hoteliere in Lausanne, Switzerland. His 25 years of restaurant and hotel industry experience include management positions at the Four Seasons in New York, Andiamo at the Las Vegas Hilton, and Delmonico's at the Riviera Hotel and Casino. Mr. Luciani frequently consults, trains, and presents to hospitality managers and employees.

A Historical Overview of Service

Outline

Key Terms

Taberna vinaria

Thermopolium

Apicius

Catherine de Medici

Guilds

Learning Objectives

After reading this chapter, you should be able to:

- Describe the importance of excellent service to a successful operation.

- Provide a historical overview of service.

- Explain how haute cuisine developed, and how it influenced modern service.

Introduction

Excellent service is vital to the success of every foodservice operation. Many operations fail not because the food or atmosphere are inadequate, but because the service fails to please guests. The National Restaurant Association has reported that 49 percent of all customer complaints involve service, compared to 12 percent for food, 11 percent for atmosphere or environment, and 28 percent for other reasons.

Excellent service depends on excellent, professional servers who not only know their jobs and perform them well, but understand their guests and how to best meet and exceed their needs. This includes the ability to work with others as a team to deliver great service, and the attitude to approach the job as a professional.

Serving is not an easy job. It requires hard work, time to learn to do it well, and a commitment to serving people.

The Age of Service

Not long ago, the economies of the world's most advanced nations were based largely on the production of goods. This is no longer true. The economies of most of the world's advanced nations now realize the rendering of service is very important, so much so that we say we are in the "Age of Service."

Serving food and beverages is a significant part of a huge and profitable industry in the dominant service sector. The National Restaurant Association estimates that yearly foodservice sales in the nearly quarter-million eating and drinking places in the United States were nearly $290 billion in the mid-1990s, nearly 5 percent of the U.S. gross domestic product. (See **Exhibit 1.1**.) The foodservice industry employs more than 9 million people. Employment is expected to exceed 12 million by the year 2005. Nearly one-half of all the adults living in the United States eat out at least once a day.

As the economy changes, people are finding that work in the service sector offers good and permanent opportunities. The foodservice industry is essential to this economy and will continue to grow, probably at a greater rate than many other service industries. Joining the service staff in the foodservice industry can provide a permanent position that pays well and gives adequate job benefits.

Exhibit 1.1—Foodservice Industry % Change in Dollar Sales (National Restaurant Association Report, 1995)

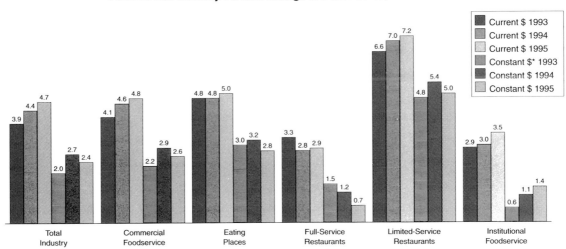

Foodservice Industry Percent Change in Dollar Sales, 1993–1995

Foodservice industry sales include all meals and snack sales at commercial restaurants, institutional organizations, hotel/motel restaurants and food contractors.
Commercial foodservice group sales include all meal and snack sales at commercial restaurants, food contractor outlets, hotel/motel restaurants and retail host restaurants.

*Represents change in sales adjusted for inflation.
Note: 1994 and 1995 figures are projected.
Source: National Restaurant Association

The foodservice industry is thriving, and highly competitive. What differentiates one foodservice establishment from another? It is often a distinctive and excellent reputation for service. Food services have found that price wars to meet competition usually do not work, but raising the level of service can be highly effective in rising above competition. People who dine out are much more service-sensitive than they used to be and will often select where they will dine by the level of service given. Food services have found that it costs very little more to provide good service rather than poor service.

Service: A Total Concept

The Meaning of Service

What do we mean by *service*? It is more than taking orders, placing down food and beverages, and clearing up after a meal. It is the act of providing customers with a wide range of meal-related benefits and experiences. Service is what makes people feel good about spending their money to eat out.

Serving should not be looked upon as menial. Too frequently servers downgrade their work. This is because they fail to understand what their task is and do not realize that serving can be professional work.

But, is serving a profession? The answer is yes. One definition of a profession is qualified persons in one specific occupation or field. We can be more precise if we add the phrase "serving the needs of others," to then say qualified persons of one specific occupation or field serving the needs of others.

Many professional people have positions that require them to serve others. A doctor serves the sick. A religious leader serves those in need of spiritual guidance. A dietitian helps others to select healthful foods. In many cultures, teaching is a highly respected service profession. Food servers meet the needs of others by serving their needs. Thus, those who serve food and drink are professionals in that they are a large body of qualified people working in one occupation, serving others' needs.

Being a professional brings on responsibilities. Professional people are supposed to meet the highest standards of moral and ethical behavior. They are expected to treat others in a professional manner. Those who serve should be proud of their work. Servers who approach their jobs professionally and are proud to serve others enhance the industry as well as their own careers. Mastering the art of service builds pride and self-esteem, and opens up a world of career opportunities.

The Tradition of Hospitality

Hospitality encompasses two important concepts: Guests should always be made to feel welcome and wanted, and all efforts should be made to see that no guest comes to any harm. These are ancient rules of custom in nearly every culture.

Many ancient peoples formalized ways in which guests were to be received when they came to one's home. An old Irish custom was to offer a pinch of salt and a small glass of wine when guests came to visit, both wine and salt being precious commodities. In ancient times, Jewish people greeted their guests by bathing their feet and rubbing them with fragrant oil. The Chinese offered special foods and drink to guests.

Another social rule that developed many years ago was that when guests were in the premises they should be protected from any harm. The concept of **sanctuary** was especially important to the early Christians, whose churches, monasteries, and convents were recognized as places of protection even from government or royal authorities. This feeling of sanctuary strongly influenced the rules of how guests should be treated at inns and restaurants. This old European value has evolved into modern laws holding innkeepers especially liable for the safety of guests. The concept has been extended, to restaurants and other hospitality operations.

Meeting and Exceeding Guest Expectations

Service has become the single most influential factor in customers' decisions when eating out. Great service gives operations a competitive edge, and keeps people coming back. A good server must learn how to read each guest to determine how to meet particular needs, and how to exceed guests' expectations.

©Culver Pictures/PNI.

Good servers do three things well: They pay close attention to detail, they work efficiently, and they are consistent even when a dozen things go wrong and threaten their demeanor. They seek the rewards—good tips, higher wages, recognition from their peers and employers—of focusing completely on the details of their work. To be efficient, servers need not kill themselves with hard work, or be rude or abrupt with customers. Instead, they must learn to plan and organize to make the best use of their time by doing the following.

- Set up work stations carefully at the start of a shift so all supplies are available.
- Replace supplies before they run out.
- Don't walk from one area to another empty-handed if there is something to carry.
- Combine trips.
- Stay organized.
- Follow the most efficient routine.
- Save steps whenever possible.
- Prepare for busy times.
- Stay on top of the job during slower times.

Today's guests are quite sophisticated. They expect good service, so the challenge is to impress them by exceeding their expectations. To do this, servers must:

- Focus completely on customers.

- Show a sense of urgency.

- Acknowledge, greet, and say good-bye to every customer with whom they come in contact.

Good servers also must anticipate guests' needs, and try to accommodate them before they think to ask. This means watching and listening to customers carefully for clues to what their needs might be, doing whatever is reasonably possible to please them, and thinking creatively when serving customers. For example:

- If a customer is standing at a quick-service operation's counter staring at the menu, a server should suggest several items, or ask if the customer has any particular questions about the menu.

- If guests in a full-service operation slow down, pause, and look around the dining room as they and their host(ess) approach a table, the host(ess) should ask if that table will be all right.

- If customers come into a quick-service restaurant with a small child but do not order food for the child, the server should ask whether they want an extra set of utensils or any appropriate children's items (coloring place mats, etc.).

- Any time a customer is looking around confusedly, a server should ask whether they need help finding something.

- A server whose customers are writing on a napkin should ask if they would like some paper.

©Christie's Images/PNI.

A Historical Overview of Service

The growth of service in food establishments is not well documented, especially in its early stages. What it was and how it grew must be gleaned from brief references in literature. In *The Canterbury Tales*, Geoffrey Chaucer writes of a nun wiping her lips daintily with a napkin, from which we can infer that 14th-century Englishwomen used napkins. Another source of how service developed is to note events or practices among people of the times. Thus, from evidence about the nearly one hundred different dishes served at a formal 17th-century dinner in France, and the elegant tableware used, we infer that elaborate service must have marked these royal events.

Beginnings

Ancient Times

Early people ate largely for survival. There was little ceremony involved. With the discovery of fire, some foods were cooked. Clay was used to make dishware and other utensils that could hold food while it was cooked over the fire. Thus, the diet changed from raw foods to stewed and roasted meat, cooked seeds, vegetables, and other items. Many of these ancient cooking pieces have been discovered. We find that the earliest pieces are crudely made, but gradual improvements were made in the clay mixtures used, and their design. People enlarged and perfected the kinds of ware used and began to make pieces from which to eat and drink. They found out how to color and glaze this ware. In some cases, ladles and cooking spoons were made. These improvements were undoubtedly a boost to their level of service.

After humans moved from caves and built dwellings, fireplace cooking developed. This was an advancement in cooking technique but service remained crude and rudimentary. Excavations in the Orkney Islands near Denmark show that around 10,000 B.C.E. (before the current era), people built their dwellings around a common kitchen and cooked their food and ate together as a communal group. There is no evidence of eating utensils.

Diggings from somewhat later times in the Mohenjo-Daro region in modern Pakistan reveal the existence of restaurant-type units where the public went to dine. The ancient Chinese also had restaurants that served food and drink in fine pottery and porcelain dishware. It is thought that the Chinese have used chopsticks since 6,000 B.C.E. It was not until six or seven thousand years later that the knife, fork, and spoon, as a place setting, were developed somewhere in southwest Asia.

Greek and Roman Times

The Romans had small eating and drinking establishments called *taberna vinaria*, from which we get the word *tavern*. These *tabernas* were so popular in the third century B.C.E. that they were found tucked away into every corner of every large city. Ruins in the volcanically preserved city of Pompeii tell us that diners could eat and drink in restaurants that featured stone counters outside and stone benches and tables inside. Cooked food was kept warm in *thermopoliums*, stone counters with holes in them to keep food warm. (See **Exhibit 1.2**.) These tabernas dispensed a significant amount of wine. Huge stone jars contained the wine, which was preserved by pouring oil over the top to prevent air from contacting it and turning it into vinegar.

©Culver Pictures/PNI.

Exhibit 1.2—*Thermopoliums* **kept food warm in Ancient Roman towns like Pompeii.**

There are many accounts of pagan feasts in honor of the gods of ancient times. From the vast amount of food and drink prepared and the numbers that attended, a fairly high level of catering service must have been developed. Greeks held feasts and celebrations that lasted for several days. Their feast to Dionysus, the god of wine, was one of great rejoicing, revelry, and excitement.

The ancient Romans rivaled the Greeks in their use of elaborate public feasts. One Roman emperor bankrupted the state treasury by stealing money for his private feasts. The wealthy and influential also gave sumptuous private feasts for large numbers of their friends, spending huge sums on them. The service was elaborate. A typical banquet had four courses. The first was called mensa prima, the second mensa secunda. Many different kinds of rare and exotic foods from all over the Roman Empire were elegantly served. Tableware included beautiful glassware and ceramic and metal dishware. Hosts vied with each other to see who could put on the most elaborate banquets. One Roman named Apicius spent so much money on one that he bankrupted himself and committed suicide. He wrote the first cookbook that we know of, and today his recipes are still used in many food services.

Street Vending

Street vending service also developed early. Wall paintings in ancient Egyptian tombs show vendors selling food in markets where people ate standing in the street. Vending and street eating were also common in ancient China, when a vendor would come down a lonely, dark street at night crying out his menu. People would come out of their homes and make a purchase. Vending of this kind is still done in many countries.

Service in Inns and Taverns

Inns and taverns were established in many ancient countries to care for land and sea travelers. China and India had laws regulating inns. Another law in China required monasteries to provide care for travelers. The writings of the Italian 13th-century explorer Marco Polo describe his stays at Chinese inns and monasteries. The ruins of some of these inns are still to be found in the dry desert areas along the ancient Silk Route. Wherever enough travel developed along these ancient routes, a hostel or inn was certain to be found.

For the most part these units offered limited service. Beds and space for travelers' animals was provided; some offered food and libations. Usually servers served the food, but in others, food was prepared by travelers or travelers' servants.

The Dignification of Dining

Ancient peoples also did much to dignify and formalize food service. Instead of just making eating an exercise in gaining sustenance, they began to attach to it philosophic or symbolic meanings, so that in the act of eating they were expressing a feeling or belief. These practices greatly influenced the kind of food served, how it was prepared, and how it was served. These customs arose around the concepts of well-being, religion, entertainment, and social reasons. Very few cultures failed in one way or another to develop such practices in these areas.

Well-Being

A number of cultures selected certain foods to eat primarily for health or sanitary reasons. Some beliefs were so firmly held that even though the results were not positive, the practices were continued with religious zeal.

However, many food remedies were effective. Today, we know much more about the need for a nutritional diet. We have the benefit of scientific knowledge, but in ancient times people learned by trial and error what one should serve and it is surprising how well some cultures did in achieving a diet that led to better health.

The Chinese led the way in establishing rules for achieving good health through food. To the Chinese, food was medicine, and medicine was food, since both nourished the body. Confucius established strict rules as to what foods to serve, how to combine foods, the amount to serve, and when foods should be eaten. About 2,800 years ago, the Chinese emperor Shennung wrote a cookbook, the *Hon-Zu*, that is still used. The art of eating well has been part of many Asian cultures for millennia.

The Chinese, and many other cultures, believed that when one ate certain foods, one took on the characteristics of the source of the food; eating tiger meat could make one fierce and aggressive, or eating an eye or a liver would make these organs in the body stronger. Even today, cultures of the world believe that the service of certain foods can bring about desirable results. Today, for example, it is the custom of many southern people in this country to eat black-eyed peas on New Year's Day to ensure good luck for the rest of the year.

A number of cultures also used the service of food and drink to signify delicate social relationships such as respect, love, contempt, devotion, or other feelings. The service of cold noodles to a guest in China indicates a lack of warmth for the relationship but, if served warm, indicates deep respect. If a Hopi Indian woman wanted to indicate a romantic interest in a man, she would give him two small pieces from a maiden's cake, made of blue corn meal and filled with boiled meat. To show matrimonial interest, she placed this on a plate of blue corn flat bread outside the door of the man's door. If he too had an interest, he took the plate inside, but if not, the plate was left outside and some family member of the girl retrieved it so she would not be embarrassed by having to remove it herself.

©Michael Newman/PhotoEdit/PNI.

Religion

The Jewish people probably developed the most complete set of religious practices using the service of food to symbolize the practices. In fact, one section of the Old Testament is given over completely to a delineation of dietary, or **kosher**, laws. Even today, people of the Jewish faith practice these customs, some of which are designed to memorialize the servitude of the Jews in Egypt and their flight to and settlement in the Holy Land. The **Seder** dinner on the first two evenings of Passover, and the eating of unleavened bread and fasting during Passover, symbolize the escape of the Jewish tribes from Egypt, a time in which they had no chance to prepare the leavened product. The serving of *haroset*, a mixture of nuts, fruits, wine, and spices, symbolizes the mortar the enslaved Jews were forced to use to build the Pharaoh's pyramids. Fresh parsley calls to mind their hopes of spring and freedom. A roasted shank bone and egg symbolize the holiday's ancient roots in a springtime sacrifice. Bitter herbs are served to symbolize the years away from home and as slaves in Egypt. Saltwater represents the people's tears during that time.

Certain rules are followed in kosher dietary laws. Pork is forbidden, only the forequarter of beef can be eaten, meat should be purged before cooking, and shellfish is not allowed. All meat must be slaughtered only by a schoichet (ordained butcher). Every seventh year wine must be poured out and returned to the ground. No fire can be built on the eve of the Sabbath, so cold food must be eaten or that kept warm in fireless cookers. In fact, so many kosher rules exist that it completely regulates dining and makes the act almost a pure practice of religious ceremony.

To a lesser degree, those of the Islamic faith have developed special food services symbolizing tenets of the creed. Ramadan, the ninth month of the Islamic year, is still spent in fasting from sunrise to sunset. Pork also is forbidden. There are a number of rules that establish customs, such as the washing of hands before dining and the proper use of the hands in eating. The Buddhist religion also established a number of practices. Eating meat is forbidden, making room for a wide number of different foods with special service requirements. On feast days, food is brought to the temples and laid at the feet of the statue of Buddha. Hindu mythology relates how Prajapati, the Lord of Creation, created ghee, or clarified butter, by rubbing butter in his hands over a fire, dropping some into the fire. He discovered the heat of the fire drove off the liquid in the butter. The people of India still consider ghee precious and ritualistically reenact Prajapati's act of creation by pouring it over a fire.

The rice farmers of Bali still practice ancient customs. The growing of rice, their basic food, has been woven into their religion. A group of farmers using water from a dam join together in a group called a tempek. This group worships and works together in the fields. They have two temples: one in the fields, and one near the dam. These temples contain their ritual calendars that indicate the time of planting, harvesting, and other activities. They have a main temple on the island's only mountain and two more near a lake in the island's center. Delegates from all tempeks meet every 210 days at the main temple to perform rituals, celebrate, and mark the intertwining of their lives with the cultivation of this staple grain.

The Christian religions have also used food and drink to symbolize theological tenets. The use of bread and wine to symbolize the body and blood of Christ is an example. Easter and Christmas are celebrated by the service of special foods and drinks.

Socializing and Recreation

Food and beverages also were used to support public entertainment and social affairs. The Greeks often used their eating places as a sort of club where they could gather and talk together about common affairs while they ate and drank. The word *colloquium* comes from the Greek word meaning to gather together to eat and drink.

Many other early cultures also used food and beverages to promote social life and entertainment. China had wine shops where people gathered to drink. Several of their greatest poets wrote their poetry there, reading it out loud to other guests. In Europe the tavern acted as a similar gathering place where people could meet socially. In the Arab countries people gathered together to dine and be entertained by acrobatic feats.

Back then, there were no electric lights, no radios, and no televisions. Newspapers, magazines, and other printed matter had not been introduced. Transportation was limited, and few ever left the area where they were born. Food was their major concern, and it was natural that they would use it as a way to enrich and extend their lives.

In ancient Egypt, meals were often simple, yet important occasions at which family, friends, neighbors, and even traveling strangers were welcome. Egyptian people ate bread, cured fish from the Nile and its tributaries, and cooked leeks and onions with meat, small game, and birds. Usually the meal was accompanied by barley wine or sweet fruit wine. Though the foods of various classes were similar, their tableware differed. Poor people ate out of glazed pottery and dishes, while the rich ate from metal dishes, used ivory and wooden spoons, and drank from glass goblets. The poor usually drank barley wine, while the rich drank fruit wine. Unlike Greeks and Romans, who often dined according to gender (a custom that continued in Europe until this century and a custom that still exists today in some south Asian nations), Egyptian women and men dined together.

In almost every ancient culture of the world, food and beverages were used as a means of worship or reverence. The Japanese tea ceremony has strong overtones of religious worship. The previous restriction against eating meat on Fridays in the Roman Catholic faith was strictly based on the fact that not eating meat was an act of reverence to Jesus Christ. The refusal to eat meat by the Buddhists and others of different faiths is a further example of ritualistic eating.

©Wood River Gallery/PNI.

In all of these customs people would pray, or otherwise indicate in their reverence their deep belief in what the service symbolized. As a result, the service of certain foods and beverages became very important by symbolizing faith and religion.

Pagan cultures also developed similar customs of reverence. Some feasts to the gods were marked by drinking wine and celebration. Often animals were sacrificed, or specially prepared and served. The use of chicken or other fowl bones and blood to foretell the future was also practiced. In Shakespeare's *Julius Caesar*, Calpurnia, Caesar's wife, begs him not to go to the senate because she has had a chicken's entrails read and the forecast was not good.

Many ancient cultures had beliefs about food and beverages that influenced their service. These ranged from strict taboos to customary practices.

The Development of European Haute Cuisine

During the Middle Ages in Europe, from the sixth to the 14th century, dining and culture in general progressed little and, in some respects, regressed. Still, inns and hostelries continued to serve travelers. In one publication from the period, we learn that inns offered three levels of service according to one's ability to pay. Monasteries took in and fed travelers on their way to the Holy Land. Public life revolved around the Church, which often sponsored community feast days. Markets offered food and drink for street consumption.

With the beginning of the Renaissance in the 15th century, the great flourishing of art, music, and architecture helped foster an environment in which dining and service, too, became more elaborate and sophisticated. Artisans and skilled tradesmen formed **guilds** to help regulate the production and sale of their goods. Several guilds involving food professionals—*Chaine de Rotessiers* (roasters of meat), *Chaine de Traiteurs* (caterers), *Chaine de Patissiers* (pastry makers)— grew in number and power until they effectively restricted their market.

During the 15th and 16th centuries, as more people ascended from poverty, the demand for better service and cuisine rose, especially in Italy. Books on social and dining etiquette appeared. In 1474, Bartolomeo de Sacchi, also known as Platina da Cremona, wrote a book on acceptable behavior while dining, dining room decoration, and good living in general. Soon after, the book *Il Cortegiano* (*The Courtier*) by Baldassare di Castiglione, became widely accepted throughout Europe as the official manual of behavior and etiquette. In 1554, Giovanni della Casa, a bishop who later was named Italy's secretary of state by Pope Paul IV, published *Il Galateo*. It was widely read and followed in its time as a guide to desirable conduct in society, and it became a classic of upperclass tastes of the European Renaissance. These last two works formed the foundation of hospitality service.

The Gastronomic Influence of Catherine de Medici

In 1533, the future king of France, Henry II, married **Catherine de Medici**, a member of one of Europe's richest and most powerful families. When she moved to France, from her home in Florence, Italy, Catherine was shocked at the inferior level of food preparation and service. Even the French court and nobility ate common stews, soup, and roasted meats. Food was brought to the table in large pots or on platters, and diners helped themselves, dishing liquids up with ladles and picking solid foods up with their hands. They ate from wooden trenchers; daggers were their only eating utensil. Liquid in the trenchers was sopped up by bread and the solids scooped up by hand. Bones and waste were thrown on the floor to be picked up by household dogs and cats.

Catherine brought a staff of master cooks and servers to her new home. Tablecloths and napkins went on the tables, and the crude dishware and trenchers were replaced with fine dishes and carved goblets made of silver and gold. She introduced French society to knives, forks, and spoons, which the Florentines had been using since they were introduced to them by a Byzantine princess in the 10th century. The foods now were sumptuous and refined, and the service was lavish and elegant.

The French court and nobility quickly adapted to the new regime, and began to imitate it. Because the use of eating utensils was so new, those who entertained did not own many, and guests were expected to bring their own.

Fortunately, the king's nephew, who would later become King Henry IV and an enthusiastic gourmet, approved heartily of his aunt Catherine's standards. When he ascended the throne, he too required the highest levels of service at court level. France's nobility became connoisseurs of fine food, drink, and service. Upperclass standards continued to rise until formal dining reached lavish and elegant levels during the reigns of Louis XIII to XVI in the 1600s and 1700s.

The great majority of Europeans who were not members of the court, the nobility, or the privileged classes continued to eat and drink simply. They ate meals primarily at home, though inns and taverns catered to travelers and continued as gathering places for people.

The Restorante

In 1765, a Parisian named Boulanger opened the first restorante on the Rue des Poulies. Above the door was a sign in Latin reading, "Venite ad me owenes qui stomacho laboratis et ego vos restaurabo." (Come to me you whose stomachs labor and I will restore you.) Boulanger claimed the soups and breads he served were healthful, easy to digest, and could restore people's energy; hence, the name *restorante*.

©Archive Photos/PNI.

The guilds objected, claiming that only they had the right to prepare and serve food to the public. They sued Boulanger to stop him legally. Boulanger countersued and started a campaign to gain publicity. He had friends in high places that supported him. Soon he made his case a celebrated cause, even getting the Assembly and King Louis XV into the controversy. Boulanger won his suit. He protected his right to compete with the guilds, and opened the door for others to start similar operations. Soon *restorantes* opened in Paris and other cities in Europe, and the foodservice industry began.

Coffee was introduced to Europe in the 17th century. This brought about the development of the coffee house where coffee was served along with other beverages and some light food. Coffee houses became popular as social gathering places for local people and acted as places where people could discuss common affairs and gain the latest news. They quickly spread all over Europe.

The **French Revolution** (1789–1799) ended the rule of the kings. Many of noble, wealthy, and influential people were killed or fled France. A new class arose, composed of artisans, capitalists, merchants, and intellectuals. This new middle class began patronizing restaurants, and the public demand for high-quality food, drink, and service increased. At the same time, many highly skilled cooks and servers who previously had served the upperclass found jobs in the new foodservice industry.

By 1805, only six years after the Revolution, 15 fine-dining restaurants could be found in the area of the Palais Royal alone, serving the nouveau riche (new rich) the finest food with the best service.

Discriminating Gourmets

As this new class grew in stature, a group of discriminating gourmets appeared and a number of them began to write about the art of fine dining. The French statesman Brillat-Savarin wrote *The Physiology of Taste*. Gimrod de la Reyniere edited the first gourmet magazine. Vicomte de Chateaubriand wrote many authoritative works on fine dining, and Alexandre Dumas père (father, or senior) compiled his classic *Grand Dictionaire de Cuisine*.

At the same time a group of chefs developed who also were interested in a high level of cuisine and service. The first of these was Marie-Antoine **Carême**, who trained a large number of very famous chefs to follow him and continue his high level of food service. They not only invented new dishes and new service, but also established rules on what foods should be served together, when they should be served during the meal, and the manner of service. Thus, it was Carême who first said that a heavy meal should be accompanied by a light soup such as a consommé, and a light meal should be accompanied by a heavy soup, such as a hearty lentil purée. It was Grimrod de la Reyniere who later voted his approval by writing, "A meal should begin with a soup that, like the prelude to an opera or a porch to a house, gives promise of what is to follow."

The Growth of Service in Modern Times

The development of service after 1900 revolves around the tremendous growth of the foodservice industry—a direct result of increased industrialization, mobility, and disposable income. Today one-fourth of all meals eaten in a day are consumed away from home. This represents 42 percent of the total dollars Americans spend for food and drink. As the century comes to a close, people are eating out often, and demanding high-quality, yet increasingly casual service.

The Rise of Hotels

Greater mobility led to the growth of hotels and motels, which in turn affected food service. Luxury hotels were built to serve affluent patrons. One of the first of these was Low's Grand Hotel, built in London in 1774. It had more than 100 rooms and extensive stables for horses and carriages. It soon had many imitators throughout England and Europe. Tremont House, which opened in Boston, Massachusetts, in 1829, was the first luxury hotel in the United States. The four-story building had 170 rooms with two bathrooms on each floor with running water. For the first time, guests could stay in their own rooms with their own key, all for $2 a night.

As railroads developed, hotels sprang up in every place with enough patronage to support them. New York City had eight in 1818; in 1846 there were over 100. Chicago had over 150.

The marriage of fine hotels, fine dining, and fine service culminated in the partnership of César **Ritz**, a hotelier, and Auguste **Escoffier**, one of history's greatest chefs. Ritz oversaw the front of the house and hotel management, while Escoffier saw to the kitchen and dining services. They made an unmatchable team; both had the highest standards. Ritz strove for elegant and luxurious service and spared nothing for the comfort and enjoyment of guests. Escoffier adapted and simplified the elaborate classic menus of his time to highlight top-quality cuisine and service. The wealthiest members of English and European society were their guests.

Ritz and Escoffier soon had many imitators. In the United States, a number of fine hotels appeared, such as New York's Astor House and Waldorf-Astoria, Chicago's Palmer House, San Francisco's Palace and St. Francis Hotels, the Silver Palace in Denver, and the Butler Hotel in Seattle. The grand balls, banquets, dinners, and social affairs held in these urban hubs displayed the finest in elaborate socializing.

By the 1990s, more than 44,000 lodging properties, with 3.1 million rooms, would be built to accommodate American travelers, diners, and trade and professional events.

Restaurants and Service in America

The first taverns in the United States were patterned after those in England. Their number increased throughout the 17th and 18th centuries, and they became an essential part of early colonists' lives. In 1656, the Massachusetts Commonwealth passed a law requiring every town to have at least one tavern. Not only did taverns provide food and drink, but they served as meeting places for people to discuss events and get the latest news. Inns also were established about the same time as taverns. They came into being largely to serve travelers and were located on the main travel routes.

The first recorded restaurant in the United States, The Exchange Buffet, a self-service, cafeteria-type operation, was built in the early 1800s opposite the present New York Stock Exchange. Boston's Union Oyster House, still in operation, opened in 1826. Delmonico's, Sans Souci, and Niblo's Garden were other fine eating establishments.

©Richard Pasley/Stock, Boston/PNI.

©Tom Sobolik/Star/PNI.

In the 19th century, dining out was restricted largely to the wealthy and to travelers. Around 1900, as the United States industrialized, workers began eating away from home more often. Cafeterias and lunch counters sprang up to serve both blue-collar laborers and white-collar professionals. As cities grew, shoppers and others were customers for cafes, coffee shops, family restaurants, and cafeterias. Institutional food service grew as well. The federal government mandated lunches in public schools in 1946. Dining out became a common experience.

Today the foodservice industry is one of this country's largest, numbering 773,017 eating places in 1993; that represents approximately $312.9 billion in sales. This places it among the top ten industries in America in numbers of units and sales. It also employs more people than any other industry, a large number of whom are servers performing an essential and important service, without which this industry—and economy—could not exist. Each year, hundreds of thousands of people are needed to fill the demand for highly qualified and well-trained servers and managers.

Chapter Summary

While the public service of food and drink began as a rather crude craft, over the centuries it grew until, by the end of the Roman Empire, it had reached fairly high professional levels. Restaurants, inns, hostelries, and other services had developed to allow the public to eat out.

Not until the 10th century did Europeans use eating utensils, although people living in the Middle East and Asia did since the 6th century B.C.E.

Since ancient times, eating and drinking have played important parts in public gatherings and celebrations. Every culture has established service and culinary customs based on their religious beliefs and physical environment.

The Renaissance ushered in an era of fine dining in Europe limited to noble and royal families. Catherine di Medici changed French eating when she became their queen, starting the growth of dining standards that reached lavish and elegant standards. Many of these standards were taken over by restorantes which started just before the French Revolution and were open to the public. Standards for fine dining also were set with the writings of a number of great French gourmets. These standards developed in France, influencing eating all over Europe and other parts of the world.

After the French Revolution, a new middle class arose with adequate incomes to eat out. Often such dining was more casual but much fine dining still existed. Restaurants thrived.

Both ancient customs and modern values dictate that guests should be treated well by their hosts, and that hosts should make every effort to see that their guests come to no harm.

Servers who approach their work professionally are able to deliver exceptional service. This entails anticipating guests' needs and wants, and exceeding their expectations.

Chapter Review

1. What part did early Christian churches, monasteries, and convents play in promoting travel and service in Europe?

2. What was a common reason for ancient Greek feasts?

3. How have religious symbols and traditions influenced food?

4. What were Catherine de Medici's contributions to French dining?

5. What was a guild? What control did a guild have?

6. Who was Boulanger? What was his contribution to food service?

7. What were some of the special contributions of the gourmet chefs of the 1800s?

8. How did industrialization contribute to the American foodservice industry?

9. How has the concept of sanctuary affected modern notions of hospitality?

10. Why is it important to exceed guests' service expectations?

The Professional Server

Outline

Key Terms

Union

Civil Rights Act

Control states

Dram shop laws

Immigration Reform and Control Act

Fair Labor Standards Act

Family and Medical Leave Act

Learning Objectives

After reading this chapter, you should be able to:

- Outline the skills and behaviors common to professional servers.

- Explain how hiring qualified employees helps an operation deliver excellent customer service.

- Describe laws that affect employees and employers.

Introduction

Many factors work together to make a server professional in the eyes of guests, colleagues, and employers: a professional, neat appearance; a positive and helpful demeanor; courtesy and tact; a high level of knowledge of food and service; the ability to accommodate many different guests with a variety of special needs; and a clear understanding of the laws and regulations that affect food service. This chapter covers these topics in detail.

Looking Professional

Servers who wear crisp, clean uniforms and are properly groomed will make a positive first impression on guests, their employers, and colleagues. This typically translates into higher tips, better shifts and table sections, and enhanced opportunities for recognition and advancement.

Styles of uniforms vary greatly from operation to operation. Many bars, family-style restaurants, and casual-theme operations feature servers in khaki pants, shorts, T-shirts, cotton button-downs, or polo shirts. Servers in upscale fine-dining restaurants often wear ties and black aprons over formal clothing. Institutional food service uniforms typically resemble traditional uniforms rather than casual street clothes. No matter what an operation's server uniform, several things are consistent. For instance, the uniform should be clean, wrinkle-free, and well-fitting. Shoes should be comfortable and sturdy enough to withstand hours of standing, walking, and direct contact with slippery surfaces.

Servers' hair should always be tied up or back if it is longer than collar-length, and their hands and fingernails should be clean and well groomed.

SOME RULES FOR GOOD PERSONAL HYGIENE

While personal hygiene may be a sensitive subject, it is vital to food safety. Illness can be spread by almost every part of the human body. Your employees should:

■ Wash hair and bathe daily.

■ Wear clean clothing on the job. Work clothes should be worn only on the job, not for personal use. If unable to change clothes at the restaurant, employees should not make any stops on the way to work.

■ Wear comfortable closed-toe shoes. Never wear platform, high heeled, absorbent-soled, or open-toed shoes.

■ Wear hair restraints. These are required by local, state, or federal health codes. Nets, hats, and caps may be used.

HAND CARE

Basic hand care includes:

■ Keeping nails short and clean. Not wearing fingernail polish or artificial nails.

■ Not touching hair, clothes, or skin—especially sores, cuts, or infections.

■ Covering all cuts and sores with bandages and plastic gloves if preparing food.

Train employees *never* to:

■ Stack plates to carry several of them at one time—their hands may touch the food.

■ Handle place settings or food without washing their hands after they have cleared tables or bussed dirty dishes.

■ Touch the insides of glasses or the eating surfaces of tableware.

If servers carry pens, order pads, corkscrews, or matches, they should be kept in pockets below the waist to prevent them from dropping. Never put a pen or pencil in your mouth or behind an ear. Extra uniforms should be on hand for emergencies. Jewelry should be limited to post earrings, rings, and watches since anything dangling could end up on or near guests' food.

The Demeanor and Attitude of Successful Servers

A pleasant demeanor and positive attitude have as much or more to do with success in serving as knowing how to do the work. In fact, a survey of managers to find out the cause of servers' failure on the job indicated that only 10 percent were discharged because they did not know how to do the job. The other 90 percent were discharged largely because of personal traits or negative attitudes.

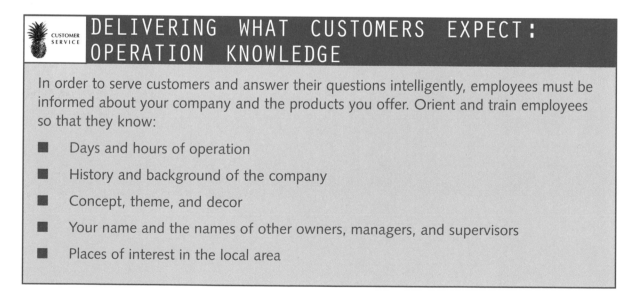

DELIVERING WHAT CUSTOMERS EXPECT: OPERATION KNOWLEDGE

In order to serve customers and answer their questions intelligently, employees must be informed about your company and the products you offer. Orient and train employees so that they know:

- Days and hours of operation
- History and background of the company
- Concept, theme, and decor
- Your name and the names of other owners, managers, and supervisors
- Places of interest in the local area

Maintaining a Positive Attitude

One of the most important personality traits a good server must possess is a positive attitude toward work, colleagues, and serving the public. Good servers believe they can deliver and try not to dwell on failures, but correct their mistakes and learn from them.

Having a positive attitude toward work allows the server to make progress in learning and to develop increased proficiency. Setting small goals and achieving them builds confidence. After this, more difficult goals appear within reach. It is important to be prepared to take advantage of opportunities when they appear.

Making an earnest effort to be friendly with guests and please them brings with it both financial and professional rewards. Servers should try to develop challenges for themselves, such as trying to win over difficult guests. **Exhibit 2.1** contains a list of the fundamentals of professional service.

Exhibit 2.1—Professional Service Fundamentals

Adhering to these standards ensures professionalism in manner and service.

- Be sure your personal style matches the style (formal, casual) of the operation.

- Don't initiate a superfluous conversation unless at the request of the guest.

- Humor can be positive and pleasing to the guest when properly applied.

- Timing is of crucial essence. For instance, if a guest orders a glass of wine served with the entree, and it comes even five minutes after, it can annoy the guest.

- Little things mean a lot. Everything that can be done to make a guest's experience more enjoyable, comfortable, and easy will always be appreciated.

- In a family restaurant, provide activities for children, such as crayons or puzzles, so parents can enjoy their meal.

- Always present the check face down, so that only the host(ess) can see the total.

- When saying good-bye, if the guest should extend the hand, the server should offer a firm handshake.

- Maintain proper eye contact, which is a sign of attentiveness and sincerity, at all times.

- Place dishes on the table gently.

- Guests should never be hurried, and should never be given the impression that they are being rushed so that others can be accommodated.

- In handling china and glassware, never touch the top or inside of a glass, or the surface or edges of a plate.

- To be a team player is a must. Help other servers whenever possible.

Servers should never talk about personal problems or inappropriate topics with guests or within guests' hearing. They should never complain in the dining room about the lack or paucity of a tip. In fact, it is not the server who should be the focus of attention at all, but the guest. While this separation is certainly a challenge, support from colleagues and good cooperation can go far in keeping servers focused on their guests and giving great service.

Courtesy

In life, courtesy means being polite, gracious, and considerate towards others. In food service, it means putting the guest's needs before one's own. Respect for others and a willingness to help are key. Courtesy should be automatic and natural. It is displayed through words and actions. Being courteous does not mean being servile or fawning. Good servers are professionally courteous, showing a serious regard for their work. Even difficult guests, when treated courteously, will return the favor. The few who don't are rare, and servers should try to meet their needs and not take their mistreatment personally.

Tact

Tact is the art of saying and doing the right thing, using the right words at the right time. It is also an intuitive sense of what to do or say in order to maintain good relations and avoid offending guests. Behaving tactfully might be remembering and using guests' names, using diplomacy in adverse circumstances, or asking a guest to take a phone call when it is necessary to tell her that her credit card has been rejected. Being tactful means handling sensitive situations so that everyone involved is left with their dignity intact.

Sincerity and Honesty

Sincerity and honesty are shown by behaving naturally, and not in a forced or phony way, toward guests. A forced smile and "canned" lines ("Have a nice day") are obvious clues to insincerity. Being pleasant while serving is really all that is needed.

Being frank and telling the truth are important. Servers who make mistakes should simply admit the mistake and correct it as quickly as possible. Guests will appreciate the forthrightness and the effort. Excellent service will be easy to perform if servers follow the tips in **Exhibit 2.2**. And in **Exhibit 2.3** are listed some helpful phrases which servers can use.

Exhibit 2.2—Excellent Service on the Job

Providing excellent service is a good way to establish loyal customers.

- Do not correct guests if they mispronounce item names.
- Anticipation is a fundamental component of service. Guests should not have to ask for refills on coffee, water, etc.

continues

- Even the most helpful service, given with improper timing, can be perceived as poor service.

- Only talk about yourself when asked. Guests are the celebrity at the meal.

- Never allow your emotions to get the best of you. The service you give must remain consistent and professional, especially when dealing with difficult guests.

- If a napkin or piece of flatware falls on the floor, replace it immediately with a fresh one.

- Before clearing something, ask if the guest is finished.

- Describe items in an appetizing manner, such as, "Our special, Southern Fried Chicken, comes from an old southern recipe using special herbs and spices; it is crisp outside and moist and tender inside."

- Too much zeal in serving can bother guests. This usually discourages tips. Service that brings in tips is pleasant, effective, and unobtrusive.

- Do not stand nearby when a guest is paying the bill. Most guests want privacy when figuring out a tip or counting out money. They may wish to discuss the tip without the server being present.

Exhibit 2.3—Helpful Phrases

Pleasant, courteous phrases are always appropriate when dealing with customers.

- Good evening (morning, afternoon) and welcome to _____.

- My name is _____, and I'll be your server this evening (morning, afternoon). If there is anything that I can get you or do for you, please let me know.

- May I take your order now, or would you prefer a little extra time to go over the menu?

- Do you have any questions about our menu?

- May I suggest a wine to complement your entree (or coffee with your dessert)?

- How is everything? Is the _____ done to your order?

- Thank you. It's been a pleasure serving you. We look forward to your return.

- I hope to see you again soon, Mr. _____ or Ms. _____.

Camaraderie

Camaraderie is the ability to get along with people. When team relationships falter, guests suffer. No matter how you serve guests in the foodservice industry, your ability to work with others to serve them well will help you move ahead professionally.

Learning Skills

Professional servers must learn continually throughout their jobs or careers. Learning and training in service skills is accomplished in several ways: through videotapes, study courses, computer programs, CD-ROM programs, simulations, training sessions, and other servers. No matter what the method, both the trainer and server are responsible for seeing that learning takes place and is put into practice on the job.

Product Knowledge

Just as a doctor knows the human body and the mechanic knows cars, servers must know about the products they serve. If a guest asks about a menu item, the server should provide all possible answers. Servers should study their menu to know how items are prepared and what they contain, and know all specials before a shift begins. This increases a server's opportunities for suggestive selling and increasing check averages and tips.

CUSTOMER SERVICE — DELIVERING WHAT CUSTOMERS EXPECT: MENU KNOWLEDGE

In addition to basic knowledge of the operation, all employees should know something about the menu. Encourage your employees to taste the food, and train them to know ingredients and methods of preparation. How much each person needs to know depends on his or her position, but train all of your employees to know something about:

- What items are on the menu
- Signature items
- Promotional items or specials
- Preparation methods of menu items
- Their favorite items

Suggestive Selling

Suggestive selling involves the extremely important role your servers have in suggesting items to guests, selling individual menu items, and increasing check averages and tips. Suggestive selling involves offering all of your guests the full range of products and services available in your operation. The more guests know about what menu items are available, how they are prepared, and why your operation is pleased to offer them, the more likely they are to enjoy every aspect of their meal. Suggesting menu items benefits your servers in increased tips, guests in increased enjoyment, and the operation in increased profits.

A knowledge of menu terminology is also essential. The best servers not only know what menu terms mean but also the explanation of the terminology. Knowing menu items can be a matter of pride and accomplishment in the profession of serving.

Inexperienced servers may not know many menu terms. The way to learn is to take a menu and ask the chef or manager to explain what is unfamiliar. Usually, in the line up session before the meal, servers are told what menu items are.

Organization

Organizing one's own work and time is essential. A disorganized server will have trouble with timing, such as knowing when to take an order, pick up items, or present the check. Disorganization breeds a frantic pace, tension and nervousness, and frequent attempts to catch up. Good servers have a rhythm.

Good organization will also help you anticipate guests' needs. Good observation is one of the most important factors in organization and timing.

Finding Work

Finding a good server position is not always easy. While there may be many job openings, it may be a challenge to find one that is a good match with your talents and availability. Word-of-mouth is one good way to find a position. This gives the server first-hand information from someone who knows the operation. Other sources of job openings include help-wanted advertisements, union headquarters, employment agencies, and just walking into operations.

The Interview

Remember when going to an interview to present yourself as a desirable employee. Do not brag but present yourself in a positive manner. Precisely explain your qualifications. Remember that body language tells the interviewer a lot. Display your confidence. Show the interviewer the same professional

behavior you will use on the job. Dress and grooming should be neat, simple, and appropriate to the operation.

As a manager, hold interviews only when you are seriously considering an individual for a position. Interviews are designed to gather information about skills, personality, and job knowledge. Often a review of the application leads to interview questions. A broad amount of information is obtained by asking open-ended questions, such as, "You have worked at several places during the last year. Why did you leave each one of them?" This can lead to further questions which might give some revealing information about the applicant. A server may go through a preliminary interview with the manager or supervisor of the operation. The supervisor who is to immediately oversee the work of the server should be the final interviewer.

Interviewing

As a manager in the competitive foodservice industry, you will have the greatest success only if you actively strive to hire and retain the most qualified, talented, and motivated employees. The process begins with screening the most qualified and appropriate job applicants, identifying outstanding job candidates through effective interviewing, and selecting employees who are likely to remain and develop within your operation.

Most interviews will be divided into four parts: 1) preparation; 2) the interview; 3) ending the interview; and 4) evaluation. The key to conducting effective employee interviews is to plan your part of the interview in advance. By preparing interview questions and structuring the interview's direction before the meeting, you are much more assured of getting valuable information about the candidate than by simply "winging it."

Most operations will find it helpful to establish **job specification**s, which are written definitions of the requirements of the job and the person that should be hired to fill these requirements. It takes a person who knows the job and what it entails to write these. The job specification lists knowledge, skills and abilities, work experience, and education and training. Write specifications based on: the information needed to perform job duties; the ability to perform a task, or behave in a certain way; and any specific skills, if applicable.

Another item that should be done before interviewing is to set up questions to use to get the information needed to judge the candidates' suitability for the job. These should cover the areas of: 1) education; 2) motivation; 3) ability to work with others; and 4) relevant personal characteristics. See **Exhibit 2.4** for some examples of questions that might be asked in these four areas. **Open-ended** questions require more than a yes or no answer, and they encourage candidates to talk about themselves and their experiences. By asking open-ended questions the interviewer gets a chance to get much more desirable information about the candidate that might be helpful in estimating the candidate's suitability for the position. Listen carefully to the candidate's answers. In fact, you should spend most of your time in the actual interview listening to the candidate.

Exhibit 2.4—Some Open Questions to Ask in Interviewing

Education

1. Who was your favorite teacher? Why did you like her or him?

2. What courses did you take? In which did you excel?

3. In what extracurricular activities did you participate?

Motivation

1. I see you have had previous experience as a server at _____. What were your job responsibilities? Did you enjoy doing them?

2. What are your professional goals? What do you want to be doing one year from now? Five years from now?

3. What do you expect from supervisors? From co-workers? From buspersons or others who work for you?

4. What did you like about your contacts with the kitchen crew? What did you not like?

Ability to Work with Others

1. What advantages do you see in working with others?

2. Describe some unpleasant experiences you have had with co-workers and how you handled them.

3. If you were given an inexperienced person to work with and train, what would you do to help this person learn, while at the same time doing the required work?

Relevant Personal Characteristics

1. When you are a guest in a restaurant, what type of service do you expect?

3. If you saw a co-worker stealing, what would you do?

There are other things one needs to do in planning for the interview. The room in which the interview occurs should be private, orderly, and unintimidating. Arrange to sit next to the applicant. Do not have anything such as a desk or table between you and the candidate, in order to help put the candidate at ease. Have materials ready to give the applicant. Notify others who will be interviewing the person. Make arrangements so you will not be interrupted during the interview. Have a note pad handy to use in taking notes.

One must remember in conducting the interview that the interviewer is also under scrutiny, and that the applicant is often making an evaluation of whether he or she wants to work there. The interviewer should greet the applicant warmly and be pleasant during the entire interview. Body language, such as facing the person who is talking and maintaining eye contact, also should be positive. Listen actively by nodding, maintaining eye contact, asking questions, and at times repeating what the candidate is saying. The only way you are going to find out what you want to know is to listen and hear it from the candidate. Of course, there are other factors to note such as dress, the way the candidate conducts himself, and so forth, but the main source is through what the candidate says. Short comments, such as "Yes, that's good" or "I see," also indicate active listening. Don't expect spontaneous answers. Let the candidate think out answers before giving them.

One should not mislead candidates nor make false promises. State frankly what is good and what is bad about the position. It is much better to be honest with the candidate about the hard parts of the job.

Before ending the interview, give the candidate a tour of the facility, explaining things that are of interest about the job. Introduce the candidate to others in the operation. Be sure to ask the applicant if there is anything more that he or she desires to know before ending the interview.

In ending the interview, thank the candidate, and indicate what will be done to inform her or him of the decision. If a date and time is given be sure to observe it. Go with candidates to the exit and wish them well. Even if the interview went poorly, still be positive and courteous. The fact that the candidate would take the time and trouble to come for an interview is worthy of polite, considerate treatment.

After the interview, quickly review your notes, adding anything that you might have wanted to jot down but did not have the time. Summarize your judgment as to the candidate's suitability for the position. Before making any decision, be sure to weigh all the facts. Today, with the shortage of labor, one is apt to make hasty decisions; avoid this. Interviewing and hiring good employees is crucial to a successful operation.

The Legal Side of Hiring

There are legal restrictions to observe in the screening, interviewing, and hiring of employees, and violators will find they face severe penalties for not observing them. Only the federal laws in this area are reviewed below, but managers and supervisors should also know and observe all state or local requirements.

The laws and their limitation affecting recruiting and hiring appear in **Exhibit 2.5**. At no time should one mention or ask the candidate to give information on any of the following:

- Race, religion, age, or gender

- Ethnic background

- Country of origin

- Former or maiden name or parents' name

- Marital status or information about spouse

- Children, child-care arrangements, present pregnancy, or future plans to become pregnant

- Credit rating or other financial information, or ownership of cars or other property

- Health

- Membership in an organization

- Voter preference

- Weight, height, or any questions relating to appearance

- Languages spoken, unless the ability to speak other languages is required of the position

- Prior arrests (convictions are legal)

The key is to ask only job-related questions. Questions such as, "Can you work nights and weekends?" and "Are you available to work overtime when needed?" are appropriate. Careful planning enables the interviewer to ask questions that are both legal and effective.

Exhibit 2.5—Federal Laws Affecting Recruiting and Hiring

Civil Rights Act (1964)	Forbids discrimination in employment on the basis of race, color, religion, or national origin; sex and pregnancy are covered in the employment section
Equal Employment Opportunity Act (1972)	Prohibits discrimination based on race, color, religion, sex, or national origin (amended Civil Rights Act of 1964)
Age Discrimination in Employment Act (1967)	Prohibits discrimination against job applicants and employees over age 40
Vietnam Era Veterans Readjustment Act (1974)	Protects Vietnam veterans from any job related discrimination
Americans with Disabilities Act (1990)	Prohibits discrimination against qualified individuals in employment; requires employers to make reasonable adjustments in facilities and practices to permit participation of disabled persons
Equal Pay Act (1963)	Requires employers to provide employees of both sexes equal pay for equal work
Fair Labor Standards Act (1938)	Establishes requirements for minimum wages, work time, overtime pay, equal pay, and child labor
Federal Insurance Contributions Act (FICA) (1937)	Source of federal payroll tax law, especially regarding Social Security
Immigration Reform and Control Act (IRCA) (1986)	Forbids employers to knowingly hire anyone not legally authorized to work in the United States

Tips

The meaning of tips—To Insure Prompt Service—is important to remember. Many servers take tips for granted, and make little or no effort to do additional work to earn one. This may be the fault of guests who tip 15 to 20 percent of the bill regardless of the quality of the service, so that servers know that extra effort is really not needed. However, with guests becoming more service and value minded, more discrimination is being shown in giving tips. Servers are expected to give good service, and in return, can expect a fair tip.

Some servers far outdistance others in earning tips. This comes through actions and words. They always welcome guests and thank them, and give the extra effort needed to please guests. **Exhibit 2.6** outlines the basics every server should know in order to please guests.

Exhibit 2.6—Basics of Good Service

Be sure to remember these service tasks when attending to guests' needs.

- Serve items from the guest's left side.

- Remove items from the guest's right side.

- Serve and remove beverages from the guest's right side.

- Serve main food items at the six o'clock position.

- Handle fine glassware by the stem.

- Keep water glasses two-thirds full.

- After filling a tray or bus pan with soiled and leftover food, cover it with a napkin.

- Serve all bar beverages with a napkin or coaster.

Tip Reporting

Tips are considered part of a server's salary, and must be reported as such to the Internal Revenue Service. When guests use a credit card and leave a tip on the sales check, a record is made. In addition, the IRS has worked out formulas for estimating cash received as tips, based on what the average server makes in various kinds of operations.

Unions

Working in a **unionized operation** means that servers are represented to management by union representatives. In a unionized operation, the employer company typically signs a contract with union representatives covering job classifications, job duties, scheduling, pay, grievance procedures, vacation time, length of work week, break times, sick leave, termination, etc. Unions charge their members dues to sustain their operation.

It is advisable for an operation's manager to have job descriptions and job specifications written before signing a labor contract if he or she does not want to forfeit that function to the union.

Union contracts often require that employees performing unsatisfactorily be warned orally and in writing a specified number of times before termination. This also happens to be a very wise management policy, since it protects managers in the case of wrongful termination lawsuit. Warnings should state a specific cause and incident description, and a description of how performance is expected to improve. Employees must be given a chance to correct their actions within a certain time period.

Grievances

Managers should make every effort to settle employee matters internally. However, if a worker has a grievance which is not satisfied, he or she can contact the union and have it take up the matter with the company management. If the union believes the company has violated a contract term, it contacts the company to settle the matter amicably. If the matter is not settled, the union may appeal to a grievance committee to hear the case and make a decision. A union contract usually contains an agreement by both the union and management that the committee's decision will be binding on both parties. If not, the employee or the company could appeal to a court for a decision. Grievance committees are limited to hearing cases that arise within a specified area.

Laws Affecting Servers

A number of laws relating to hiring and work affect both employers and servers. The most important of these follow.

Privacy Act

The **Privacy Act** of 1974 forbids employers from asking non job-related questions which might discriminate against a group of qualified job applicants. The act applies not only to interviewing potential employees but also to discussing matters with current employees.

Fair Labor Standards Act

The **Fair Labor Standards Act** of 1938 protects workers between ages 40 and 70 from discharge because of age. This act also covers teenage workers, working hours, and union activities.

Family and Medical Leave Act

The **Family and Medical Leave Act** (FMLA) of 1993 requires employers with 50 or more employees to offer up to twelve weeks of unpaid leave in any twelve-month period for any of the following reasons:

- Birth, adoption, or foster care of a child
- Care for a child, dependent, spouse, or parent with a serious health condition
- Care for the employee's own serious health condition

The Act includes some other provisions for both employer and employee that guide their relations dealing with family and medical leaves.

Civil Rights Act

Discrimination against job applicants and employees is prohibited in Title VII of the **Civil Rights Act** of 1964. The Act is administered and enforced by the **Equal Employment Opportunity Commission** (EEOC). It is unlawful to "fail or refuse to hire or discharge any individual or otherwise discriminate against any individual" on the basis of race, color, religion, sex, or national origin. The reason for not hiring an applicant must be job-related. For instance, in New York City a group of restaurants refused to hire women as captains and servers, saying that males only would be accepted by their guests. The Commission ruled against them. In some cases, foodservice operations must train applicants to become suited to the work.

The Civil Rights Act further covers wrongful discharge on the basis of age, disability, and participating in collective bargaining or union activities. The penalty to the employer may be re-employment, court costs, attorney's fees, and in some claims penalizing the employer for committing an act against public policy or outrageous employer conduct.

The Act now also covers sexual harassment on the job, which is defined as "unwelcome sexual advances, requests for sexual favors, and other verbal or physical conduct of a sexual nature. . . when:

1. Submission to such conduct is made either explicitly or implicitly a term or condition of a person's employment; or,

2. Submission to or rejection of such conduct. . . is used as the basis of employment decisions, affecting such person; or,

3. Such conduct has the purpose or effect of unreasonably interfering with a person's work performance or creating an intimidating, hostile, or offensive working environment."

A majority of cases on sexual harassment have favored the plaintiff and imposed substantial penalties on offenders. Judgments against defendant employers can be substantial.

Beverage Alcohol

Sale of alcohol is regulated by state and local laws. They generally cover licensing, permits, and how liquor may be sold. Some states called **control states** handle the sale and distribution, while others permit retail beverage establishments to purchase from prescribed dealers. They also typically address how to handle disruptive patrons. It is wise to call in the police to help handle a difficult guest.

Dram shop laws hold anyone serving alcohol liable if an intoxicated patron injures or kills a third person; hence, the equivalent term "**third-party liability**." Courts may grant large damages against establishments and servers for violating dram shop laws. Insurance companies faced with paying such penalties have raised insurance fees to cover these costs.

To reduce such liability, many jurisdictions require that those serving alcohol be trained in serving it responsibly. Responsible service includes never serving alcohol to three types of people: intoxicated patrons, minors, and known or habitual alcoholics. Detecting a minor can be difficult. An operation should have specific steps in place for verifying identification. Documents without pictures or dates should not be accepted. If in doubt, examine the documents carefully. At times, minors accompany older persons who buy them beverage alcohol. Some operations or laws may prohibit minors from entering the premises of an operation selling beverage alcohol.

Detecting intoxicated people and alcoholics also is difficult. Servers must be trained to observe and interpret guests' behavior, and monitor guests' drinking. If an intoxicated guest tries to drive away, a server or manager should call the guest a cab, or, if all else fails, call the police.

Immigration Reform and Control Act

The **Immigration Reform and Control Act** (IRCA) of 1986 makes it illegal to hire aliens not authorized to work in this country. Employers must verify citizenship at the time of hiring. If asked before hiring (during the interview), non-hired applicants can claim discrimination. There is a grandfather clause which says that aliens hired before November 7, 1986, can be kept on the job. In addition to these, some aliens may be given a permit to work in this country. Every new employer must complete a Form I-9 upon hiring and put it into the employee's file; its purpose is to verify that the new employee has submitted satisfactory proof of identity and work authorization, if the latter is required.

Americans with Disabilities Act

The Americans with Disabilities Act prohibits discrimination against people with differing levels of ability, requires reasonable accommodation for employees with disabilities, and requires public places and services to accommodate guests with disabilities. This means that not only is it good business to accommodate all guests, regardless of ability, but it is also the law.

Chapter Summary

Looking professional is key to a server giving a good first impression. Demeanor and a positive attitude are crucial. Servers must be sincere, honest, and courteous. Good teamwork helps give better service.

It is important for servers to know menu terms and how items are prepared.

Unions represent employees in dealing with management. When looking for work, word-of-mouth can be a reliable source, as are newspaper advertisements, employment agencies, and walking into operations. The art of applying and interviewing for work is very important in finding desirable positions.

Some servers are able to make more tips than others. Much of their success comes from knowing how to please guests. Tips are considered salary and must be reported as such to the IRS.

Certain laws have an influence on how servers get work and how they perform it. The Privacy Act protects candidates and employees from inappropriate inquiries. The Fair Labor Standards Act protects servers from the ages of 40 to 70 from discharge because of age and has sections governing the work of teenagers. The Family and Medical Leave Act requires employers with 50 or more employees to offer up to twelve weeks of unpaid leave in any twelve-month period for reasons related to family and personal health.

The Civil Rights Act bars discrimination against employees because of race, color, religion, sex, or national origin. The Equal Employment Opportunity Commission (EEOC) enforces the act, which also covers cases involving sexual harassment on the job.

State and local dram shop laws create third-party liability on people serving beverage alcohol. The laws hold servers responsible to third parties injured or killed by intoxicated patrons, and encourage servers to be trained in monitoring alcohol service to guests.

The Immigration Reform and Control Act holds employers responsible for verifying the legal working status of employees. With a few exceptions, illegal aliens are prohibited from working in the United States.

Chapter Review

1. In an operation that is unionized, what procedures will take place to bring a server from the first warning to a hearing before a grievance committee?

2. Is the question, "Do you have any disabilities?" appropriate to ask in a job interview?

3. What should servers do to maximize tips?

4. What constitutes sexual harassment?

5. What is third-party liability?

6. Describe a typical servers' uniform for a casual operation.

7. How does being tactful help a server when a guests' credit card has been denied?

8. What do job specifications contain?

9. Describe some ways to show you are listening actively during an interview.

10. Is it appropriate to ask a female candidate if she is planning to have a baby anytime soon?

Exceeding People's Needs

Outline

Learning Objectives

After reading this chapter, you should be able to:

- Explain the steps in resolving customer complaints in order to satisfy guests.

- Describe ways to manage service to customers with special needs.

Introduction

It is important for foodservice managers to create a work environment in which all employees are encouraged to try to satisfy and please guests. They must be trained to focus on guests at all times, and put their needs first. From the moment guests enter the operation to the time they leave, servers must make them feel comfortable, welcome, and anxious to return. By anticipating guests' needs, employees will be able to serve guests in the most efficient and effective way possible.

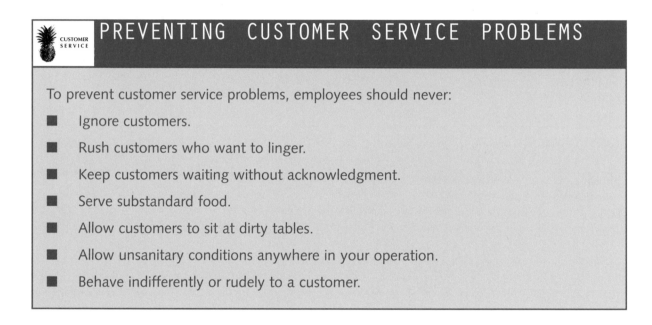

PREVENTING CUSTOMER SERVICE PROBLEMS

To prevent customer service problems, employees should never:

- Ignore customers.
- Rush customers who want to linger.
- Keep customers waiting without acknowledgment.
- Serve substandard food.
- Allow customers to sit at dirty tables.
- Allow unsanitary conditions anywhere in your operation.
- Behave indifferently or rudely to a customer.

Managing Guest Complaints

To solve a guest complaint, the server should ascertain what the complaint is, correct it if possible, and show the guest the complaint is a serious matter to the operation. This will usually satisfy most customers. Be sure to offer a sincere apology for any inconvenience. Showing sincerity, concern, and a desire to correct the problem is the right approach. The following study by the National Restaurant Association indicates that poor service accounts for the majority of restaurant customer complaints. (See **Exhibit 3.1.**)

Exhibit 3.1—From *Tableservice Restaurant Trends* (1993, by the National Restaurant Association)

Most guest complaints are about the service.

Complaint Category	Occurrence Percentage
Service	49%
Food	12%
Atmosphere	11%
Other	28%

The report advises, "The association's research suggests that consumers' levels of expectations are rising and that satisfying customers will become even more difficult for operators." Interestingly enough, the same study found, in 1995, that the service that customers receive was indeed improving. The report states that in 1994 about seven out of ten restaurant patrons were satisfied with the service they received, up from about six out of ten in 1990. The report commented, "Clearly, operational efforts at motivating and training employees are likely resulting in more satisfied customers."

Complaints as Opportunities

Some guests complain with good intentions. They may be complaining about something they think might cause others to complain and they are trying to call it to management's attention. Usually the complaint is given in a calm, friendly, and suggestive manner. Management should be extremely grateful, promise to take care of the matter, and then do so.

Do not look at a complaint as an attack on anyone in the operation or how it is managed. Complaints can be helpful in improving business. Often, managers learn about problems only from their customers. It is not easy to attract new customers, so it is never wise to lose any. Receiving a complaint is much better than having a dissatisfied guest leave and never come back. Accepting the complaint allows one the chance to correct it. The silent, dissatisfied patron is a hazard to profitable business. Unhappy guests who do not come back often discourage others from coming. While word of mouth may be the best advertisement, it also can be the most destructive. **Exhibit 3.2** lists eight of the most frequent complaints made by customers.

Exhibit 3.2—Most Frequent Complaints by Foodservice Patrons

1. Inconsistent and untimely flow of service.

2. Poor food quality and improper serving temperature.

3. Poor sanitation practices and poorly kept facilities.

continues

4. Staff not friendly and lacks product knowledge.

5. Too long of a wait to be seated, served, and presented with the check.

6. Cleanliness of service ware, utensils, and equipment not acceptable.

7. Ambiance and decor lack character.

8. Inconvenience of parking, accessibility, and location.

CUSTOMER SERVICE — DEALING WITH ANGRY GUESTS

Tips for handling angry guests:

■ Don't be defensive.

■ Don't assign blame.

■ Don't ignore the guest with the complaint.

■ Be sure not to make promises you can't keep.

■ Don't keep phone guests on hold.

■ Don't say no to a guest's request without an explanation.

Taking Action

No one appreciates being attacked by an irate customer. Servers must be trained to stay calm and composed in these situations, detached from personal negative feelings, and staying professional at all times. The rule in receiving a complaint is: Listen, accept responsibility, then resolve.

The first thing a server should do is listen to the guest and find out the real cause of the complaint. Try not to give excuses, such as, "We're short three servers tonight." Ask questions until the complaint is understood. Restate it so the guest knows you understand the complaint correctly. Then work together to find a mutually acceptable solution. An apology for having caused the guest any trouble is also in order.

Servers should try never to let a dissatisfied guest leave angry. If servers cannot solve a problem they should find someone who can. It is almost always worth it to give something to an unhappy guest, since that is an investment in future business. It might be a free item or free meal. If a guest ever becomes abusive or violent, it is best to call management, security, or the police and let them handle the situation. At all costs, try to avoid a physical confrontation. It is important to log all complaints in order to determine problem areas and avoid similar complaints in the future.

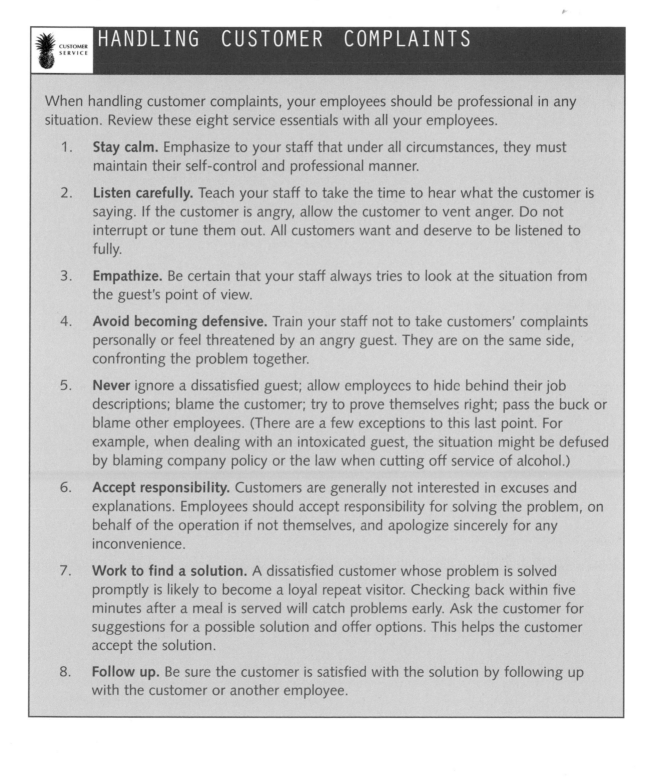

HANDLING CUSTOMER COMPLAINTS

When handling customer complaints, your employees should be professional in any situation. Review these eight service essentials with all your employees.

1. **Stay calm.** Emphasize to your staff that under all circumstances, they must maintain their self-control and professional manner.

2. **Listen carefully.** Teach your staff to take the time to hear what the customer is saying. If the customer is angry, allow the customer to vent anger. Do not interrupt or tune them out. All customers want and deserve to be listened to fully.

3. **Empathize.** Be certain that your staff always tries to look at the situation from the guest's point of view.

4. **Avoid becoming defensive.** Train your staff not to take customers' complaints personally or feel threatened by an angry guest. They are on the same side, confronting the problem together.

5. **Never** ignore a dissatisfied guest; allow employees to hide behind their job descriptions; blame the customer; try to prove themselves right; pass the buck or blame other employees. (There are a few exceptions to this last point. For example, when dealing with an intoxicated guest, the situation might be defused by blaming company policy or the law when cutting off service of alcohol.)

6. **Accept responsibility.** Customers are generally not interested in excuses and explanations. Employees should accept responsibility for solving the problem, on behalf of the operation if not themselves, and apologize sincerely for any inconvenience.

7. **Work to find a solution.** A dissatisfied customer whose problem is solved promptly is likely to become a loyal repeat visitor. Checking back within five minutes after a meal is served will catch problems early. Ask the customer for suggestions for a possible solution and offer options. This helps the customer accept the solution.

8. **Follow up.** Be sure the customer is satisfied with the solution by following up with the customer or another employee.

Why Guests Complain

The reasons guests complain are varied. All complaints are legitimate to the guest making them, and can be beneficial to managers. Complaints can be opportunities to help improve one's business.

Some complaints are simply the consequence of a guest's disposition at that time. It could be a moment of swinging moods or an evidence of poor health. Some guests complain because of unrelated stresses. If work has gone poorly that day or the guest is upset over an argument, a guest is more likely to be impatient with small errors. Personal issues may cause guests to complain about the food when it really is of good quality. Servers need to know this, and respond with calm understanding. A high degree of tolerance and patience is needed particularly in extreme cases, like when guests walk in the establishment with the premeditated intention of obtaining a free meal. Upon finishing the meal, they will act irate and dissatisfied with the quality of the food and service. In this case, servers must be trained to handle the situation patiently and professionally. Management should take steps to see that all employees are trained to handle complaints.

CUSTOMER SERVICE · SIGNS OF AN UNHAPPY GUEST

Some subtle signs that indicate a dissatisfied guest:

- Looking irritated.
- Avoiding eye contact.
- Not starting their meal after it is delivered table side.
- Asking to see a menu after their meal is delivered in order to check the menu description.
- Saying unconvincingly that everything is "okay."
- Looking around the dining room attempting to make contact with staff.
- Not finishing any part of their meal from appetizer to dessert.
- Abruptly requesting the check.

Avoiding Complaints

Many complaints can be avoided by adhering to service standards. Servers must take orders carefully and communicate thoroughly with guests. For instance, if a preparation time is going to be quite long, the guest should be told. Remember that guests want explanations, not excuses.

At the start of the meal, some guests show they might be difficult. Astute servers can sense this and must take special care to satisfy them. The server who gets the least complaints is the one who understands people and knows what pleases them.

If a complaint is likely because of something in the kitchen, the server should bring this to a cook's attention. All servers should try to correct mistakes before they make it to the guests table.

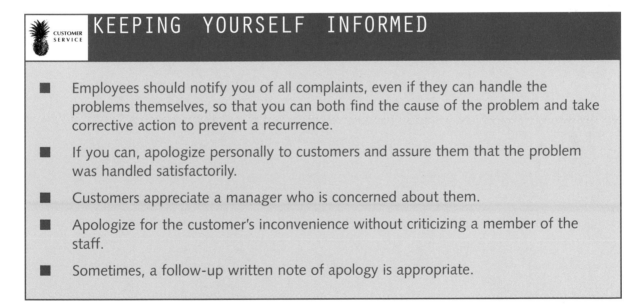

KEEPING YOURSELF INFORMED

- Employees should notify you of all complaints, even if they can handle the problems themselves, so that you can both find the cause of the problem and take corrective action to prevent a recurrence.

- If you can, apologize personally to customers and assure them that the problem was handled satisfactorily.

- Customers appreciate a manager who is concerned about them.

- Apologize for the customer's inconvenience without criticizing a member of the staff.

- Sometimes, a follow-up written note of apology is appropriate.

Serving Guests with Special Needs

Servers must know how to serve guests with special needs. No matter what that special need is, servers should look to guests themselves for solutions, and offer only as much assistance as is welcome.

People with Disabilities

A guest with a physical disability might use crutches, a cane, a walker, or a wheel chair. Often members of the party will help the guest to his or her seat; the server may only need to take the walking aid and put it away in a safe place.

Sometimes it is useful for a server to meet the guest at the front of the establishment and walk ahead, making sure the way to the table is clear and that the guest has the time needed to get seated.

Crutches and canes should be stored near the guest's table, but not in the way, or directly under the table. Armchairs are helpful because people can put their hands down on the arms and lower themselves into the seat. When the customer is ready to leave, the crutches should be brought quickly to the table.

Guests in wheel chairs may only require that the server remove the chair at the table so the guest can pull up. If the guest requests it, the server can help him or her move from the wheel chair to the chair, and store the wheel chair out of the way.

Those with a disability of the arms or hands may have special needs. Customers with some illnesses may be able to eat only with a spoon. Stemmed glassware is easily knocked over, so tumblers are preferred. Straws may be helpful. Serve hot beverages in a deep cup so any shaking of the cup does not spill the hot liquid on the guest. In operations such as a nursing home, some guests might carry eating utensils which are specially designed for them.

Communication Issues

Servers should pay careful attention to a guest who communicates unconventionally, or who nods or points to the item to indicate understanding. A small pad will be helpful for writing questions ordered, such as, "How would you like your steak cooked?" Another member of the party might give the order.

Some people simply have difficulty hearing. If the server faces the guest and speaks clearly, the guest often has no difficulty hearing. Again, the server may confirm the order by pointing to the menu, or by repeating the order back to the guest.

Guests with speech impairments might be difficult to understand. If a server has difficulty, he or she can ask, "Did I understand you to say...?" to be sure. Ask questions courteously in order not to cause the guest any embarrassment.

Guests who speak little or no English may have difficulty getting servers to understand them. Someone else on staff might help. Body language, pointing, and nodding can help. Have the guest point to items on the menu, if some ability to read English rather than speak it is shown.

People with Vision Impairment

When serving a sight-impaired guest, be sure to greet him or her each time you approach the table, so your presence is known. Offer to lead a sight-impaired guest to the table. It is often helpful if the server offers an arm, but take your cue from the guest. At the table, pull out the chair and put the hand of the person on the back of the chair. This locates the person. If necessary, some guidance may have to be provided.

Guide dogs are allowed in food services, and will usually follow a server to the table, guiding its owner. The seating is then the same, with the server now holding the dog's leash, while the guest sits. Lead the dog to a place near the table where it is out of the way, but can see its owner. Do not pet, feed, or disturb the dog. When the guest has finished the meal, help the customer up, and let the dog follow the server out.

When presenting the check to a sight-impaired guest, clearly state the total, then go over the individual prices. If using a credit card, you might want to restate the amount before the person signs the slip. The server may have to help the customer find the proper line to sign. If the guest pays with cash, state the bill denominations and coins of all change.

People with Special Dietary Needs

More and more guests dining out want foods that meet special dietary needs. This is partly because our population is aging, and as age increases special dietary needs increase, and partly because customers are becoming very health-conscious. Many food services are able to provide low-salt, low-fat, low- or high-fiber, low-cholesterol, low- or high-calorie, low- or high-protein foods. It is beyond the responsibility and mission of most food services to enforce healthful eating, but most make a variety of foods available to satisfy most needs and tastes.

Servers should know enough about each of these dietary needs to assist guests in making selections and act to meet those needs. Guests will appreciate the assistance.

A good working relationship with the chef and cooks is essential in every application of service, particularly when dealing with guests who have special dietary needs. Special requests are more and more common. Many people ask for ingredients to be omitted, such as salt, oil, butter, or cream.

Managers should reward the servers who help their guests and ensure their repeat business.

Chapter Summary

When dealing with guest complaints, the server should make sure to offer a sincere apology for any inconvenience. It is important to show sincerity, concern, and a strong desire to amend the situation.

Complaints can be helpful to your operation. Accepting a complaint from a guest allows one to correct the problem and avoid similar situations in the future. The rule in receiving a complaint is: Listen, Accept Responsibility, and Resolve. If servers cannot solve the problem, they should find someone who can. By adhering to service standards, many complaints can be avoided.

Servers must know how to serve guests with special needs. Guests using crutches, canes, and walkers may have them stored near the guests table, but not in guests' way. Guests in wheelchairs may need the server to remove the chair for them, or help move them from the wheelchair to the chair at the table.

Special utensils may be needed by some guests with a disability of the hand or arm. Some may only be able to use a spoon, or may need to drink from a tumbler with a straw to avoid spillage.

When serving guests with hearing and speech impairments, servers should speak clearly, use a notepad, or point to items on the menu to clarify what the guest wants. If the server is unclear as to what the guest wants, he should ask again.

Vision-impaired guests may need to be assisted to the table. A server may need to describe the menu to the guest, explaining each item in detail. Guide dogs should rest near the table where they can see the owner. When presenting the check to a sight-impaired guest, the server should clearly state the total and individual prices.

Servers should be aware of various dietary needs, and be able to assist guests in selecting items to meet those needs. Having a good relationship with the chef and cooks in your operation will help procure the desired meal from the kitchen.

Chapter Review

1. Describe how you would serve a guest who is deaf or hard of hearing.

2. A guest in your operation has a complaint about the slow service and poor food. How do you handle the situation?

3. Describe how you would serve a guest who is blind.

4. When receiving and accepting a complaint, what three steps should you be sure to follow?

5. A family of four has just been seated at a table in your section. The parents look tired, and the two children are crying. Should you be aware of outside circumstances that may cause the guests to complain? How can you avoid a complaint in this case?

6. Where should you place a guest's crutch, cane, or walker?

7. How do you seat a guest in a wheelchair?

8. List some ways to communicate with guests who have a speech impairment, or speak little or no English.

9. How do complaints improve business?

10. You should never give anything free to an angry customer—it only encourages them to continue behaving that way. True or false?

Service Mise en Place

Outline

Key Terms

Mise en place

Grimod de la Reynière

Silencer

Traffic sheet

Learning Objectives

After reading this chapter, you should be able to:

■ Oversee the maintenance of well-stocked and organized service stations.

■ Explain the importance of mise en place for servers.

Introduction

Many preparation steps must be completed before guests arrive to eat in a dining area. Not only must the restaurant be clean and in good order, but equipment, food, beverages, dishes, and a host of other things must be ready for use. Without such prework, service becomes chaotic and disorganized. Not only are guests dissatisfied, but servers are frustrated and tired. Preparing well for the busy period of service that follows can pay off in making the work much more efficient, easier, satisfying, and financially rewarding.

The prework that takes place both in the kitchen and guest areas before service begins is often called *mise en place*, which means in French, "put into place." It means getting everything ready for serving guests, but it also means keeping things in good order as one works. In food production it means the same thing—have everything ready, and work to continuously keep things in order. Another commonly used term is "side work."

Have a place for everything, clean up as you work, and think ahead. In both the service and production areas, good mise en place often denotes an effective worker—and a much happier one. Good mise en place makes work fun; the lack of it makes it a drudgery. Most importantly, good mise en place makes for satisfied guests.

Mise en place work is usually divided into three parts:

1) Getting ready

2) Sidework during service

3) Ending the meal

The first part consists of all work done before guests arrive, such as setting up the station, preparing ahead to meet guest needs, and arranging for service in an orderly fashion.

One group of guests may leave a table full of soiled dishes and linens, dirty ash trays, crumbs, and other things that need attention in order for a fresh, neat, correctly set table to be ready for the next group of guests. This is the second part. The last part is the departure of guests, closing the station and leaving it in good shape for the next meal. No one part is more important than the other. Each must be done correctly to achieve totally satisfactory mise en place.

Mise en place work differs for different serving situations. For breakfast, the table setting is different: perhaps a superfine sugar is put into sugar bowls rather than regular, table sugar. The condiments required might be minimal compared to those set out for lunch.

As an example of how a different situation can influence the mise en place, the type needed for counter service differs from that of a regular meal, and the mise en place needed for a buffet dinner is quite different from that of a banquet. In general, banquet service will have no part three mise en place because there is just one setting. If another banquet, party, or meal is to follow one banquet, the mise en place is different and the work to prepare starts all over. In quick-service restaurants, employees and managers must be diligent in their efforts to keep guest areas clean, since guest turnover is quite fast.

Getting Ready

Arrival

Upon arriving at a food service establishment, guests see the outside and the surrounding grounds first. This must be inviting and give a favorable impression of what is to follow. *Grimod de la Reynière*, editor of the first gourmet magazine, said, "The soup to a meal should be like a porch to a house or the overture of an opera; it should be

an inviting prelude to what is to follow." This same could be said of the entrance and grounds of an operation, no matter whether it is a drive-in, restaurant, or a boat waiting at a dock to take an evening party sailing in the bay while serving dinner.

The grounds should be clean and in good order. There should be good lighting and good security. Signs should be well lighted and visible from some distance so guests can be prepared to stop. Landscaping should be attractive, well groomed, and appear fresh. Walkways should be uncluttered and in good condition, and guests should not have to walk too far to enter. If valet parking is available, it should be prompt and courteous. Valet employees should be neatly groomed and uniformed, and should assist guests exiting their vehicles.

When entering a foodservice operation, there is always a first overall impression. This is made up of a combination of factors including color, sound, and decor. Quick-service facilities and employee cafeterias strive for a feeling of brightness and movement because they are busy places where people do not linger. Such a feeling helps contribute to higher guest turnover. In fine dining restaurants, conversely, people come to dine slowly and relax. These operations need less light and should be quieter and slower-paced.

All doors and windows should be clean and bright. The vestibule or hall should be well lighted and attractive. If there is a coatroom or area for accommodating guests' outerwear and other items, it should be near the entrance. A sign should indicate that guests are responsible for the items left, but employees should be observant of any breach of security.

LARCENY—PERSONAL PROPERTY OF PATRONS

Customers or others may victimize other patrons by stealing their personal items, vandalizing, or stealing their vehicles in the establishment's parking lot. Despite the presence of signs warning customers that they are responsible for their vehicles, the courts may hold foodservice establishments responsible for customer cars and their contents in the restaurant's parking lot.

Usually these crimes are reported to the establishment by the victims. The establishment should immediately call the police. If the victim accuses another patron of larceny, a manager should proceed very cautiously. The first step is still to call the police. Caution both the victim and the accused against violence. Ideally, the victim and the accused will wait for the police to arrive and begin an investigation. The establishment should restrain the accused only in self-defense or to protect others. Consult your attorney about the establishment's obligations in apprehending suspects.

Many operations decorate entrances with pictures, plants, or art. In full-service restaurants, the host stand should be uncluttered and out of guests' and employees' walk ways.

The effective use of indirect lighting can add depth to a dining room. Shadowing helps give a pattern and breaks up the lighting. The light, however, should not be so dark as to make it difficult for people to see as they dine or move about. Sufficient lighting should be provided at tables so guests can easily read menus.

LIGHTING

Lights are essential to any system. Employees and customers need to see to protect themselves. Exterior lights in the parking lot, near walks leading to the establishment, in stairwells, near entrances, and in vestibules should be bright enough to allow a person to read a newspaper. Exterior lights should be protected by wire cages over the bulbs, and serviced by a back-up power supply. Interior lights in hallways, storage rooms, and basements should be bright enough so that anyone hiding in them can be seen.

Lighting also serves as a deterrent to crime. Criminals do not usually want to be seen or identified, so they will avoid a well-lit establishment in favor of one that has dimmer lighting.

Burned-out lights should be replaced immediately. Because of the danger of accidents, the use of candles or burning lamps for table light and decoration must be done carefully. Check all table lighting to see it is in good working order and functioning properly.

GENERAL SAFETY AUDIT, FACILITIES, LIGHTING

Is lighting adequate in all areas? In dining areas, is the light sufficient during meal periods for servers and bussers to move safely? Can the lighting be raised to a higher level for cleaning? In the kitchen, is the lighting arranged so no shadows are cast on food preparation areas? Fluorescent lighting provides the most uniform and energy- and cost-efficient lighting for kitchens. In stairways, corridors, ramps, and pathways, is the lighting sufficient and located so no shadows are cast? Is lighting sufficient for cleaning in the storage area? Is heat-proof lighting provided over cooking areas, in vent hoods, and other highly heated areas? Are light fixtures, bulbs, and tubes shielded? Shielded fluorescent lights will not shatter into food. Are lamps and work area surfaces requiring lighting kept clean so light is fully provided and reflected? Are lights replaced before they fail or begin to flicker? A regular replacement schedule based on the life expectancy of the type of bulb used avoids damage to the fixture.

Almost all operations will have a telephone available where guests can make calls. They should be neat and clean, and phone books should be available for use. Check frequently to see that these books are in order and do not have missing pages. Good lighting is also important near telephones.

A facility may also need to make arrangements for lost and found items. This can be at the cashier's station, which is convenient for guests to locate.

The cashier's station and host or hostess desk are often located just outside the dining area. They should be kept neat and orderly so guests do not see a messy area as they pass by.

First impressions on entering the dining room should be pleasing and should give guests a welcome feeling of cleanliness and order. They should be inviting and positive. If the first impression is negative, it could linger on through the entire meal, and might affect an otherwise perfect experience.

The Dining Area

All guest areas should be neat, orderly, and inviting. Dirt or spills will spoil an otherwise perfect impression and negatively color a guest's feelings about your establishment. A dirty spot on the entryway carpet may spoil the cordial welcome the facility is trying so hard to create. That's why it's important for all employees to see that the areas for which they are responsible contribute to an overall favorable impression.

The decor of the dining area should be fitting to the ambiance and clientele of the operation. Simplicity at times can be better than too much lavish decor. In full-service operations, what guests should see is a neat, orderly grouping of tables, set with the proper flatware, decor, and glassware. (Some authorities recommend that glassware be turned upside down, while others object to this practice, saying that inverted glassware indicates the table is not ready for guests.) Candles and table decorations should blend in with or complement the decor.

Quick-service operations should have only clean, neat tables and booths. There should be a convenient place for depositing waste and trays.

Whatever the case, the arrangement of tables should be neat, not too close together, and clean. Chairs should be squarely set slightly under the table.

Before a meal begins, make an inspection to see that the dining area is ready for guests, and make any necessary corrections. Ask yourself, "Would I want to eat here?" Your servers might also ask themselves this same question. **Exhibit 4.1** is a checklist used in one operation to see if a dining room is in order.

Exhibit 4.1—Sample Dining Room Checklist

Answering "yes" to each item indicates excellent mise en place.

DINING ROOM CHECK LIST

Inspected by _____

Time and Date _____

Area _____

Temperature _____ Humidity _____ Air Quality _____

Yes	No		Yes	No	
❑	❑	All lights burning?	❑	❑	Traystands stored in proper locations?
❑	❑	Fixtures clean?			
❑	❑	Windows clean?	❑	❑	Silencer on properly?
❑	❑	Draperies properly arranged?	❑	❑	Table linen clean and pressed?
❑	❑	Furniture and decorative items clean and dusted?	❑	❑	Table cloth arranged properly?
❑	❑	Stations inspected?	❑	❑	Napkins properly folded and placed?
❑	❑	Tables level?			
❑	❑	Tables arranged properly?	❑	❑	Table lamps clean and in good working order?
❑	❑	Aisle space adequate?			
❑	❑	Chairs in good condition?*	❑	❑	Flower vases clean and filled with fresh flowers?
❑	❑	Clean?	❑	❑	Flowers nicely arranged?
❑	❑	Chairs arranged properly at the tables?	❑	❑	Covers set properly for meal or occasion?
❑	❑	Carpeting clean?	❑	❑	Dishes clean?**
❑	❑	Floors clean?	❑	❑	Flatware clean?
❑	❑	Floors polished?	❑	❑	Glassware clean?

*Check legs for splinters; seat covers for rips and stains.

**Check each setting for proper flatware and placement; check also for proper placement of dishes, etc.

Environmental Control

The lighting, air conditioning, heating, and ventilation should be comfortable. Check for drafts. Be careful to eliminate any unpleasant cooking odors in the dining area.

Managers should know the basics of mechanical adjustments so they can help maintain a comfortable environment and correct problems. A manager should always be notified of heat and humidity problems.

Humidity can cause guest and employee discomfort. When the air is too moist, it does not evaporate well from the body and this causes body heat to build up. The air in a dining room should have around 50 percent relative humidity. To check whether the dining area is too humid, just fill a glass with ice and water and set it down on a dish. If beads of moisture quickly build up on the surface, the room is too humid. Treat air to remove excess moisture as well as cool it down.

Servers should check to see that tables near cool air outlets are not blasted with cold air. Nobody wants to sit in a cold draft. Direct sunlight coming through a window quickly builds up heat. Proper window coverings can control the sun's rays. A desirable temperature in the dining area is around 70°F (21.1°C.) If the temperature is hot and dry outside, the temperature can be somewhat higher and guests will still feel comfortable. Dry air helps people feel cool even though the temperature may be higher than 70°F.

Those in charge of dining areas should understand that guests themselves build up heat and moisture from their bodies and from the air they are exhaling. It is said that every adult gives off as much heat as a 100-watt light bulb. Thus, in a well-lighted dining room filled with guests and servers, heat can build up considerably and must be monitored to ensure everyone's comfort.

Station Mise en Place

Dishes, flatware, glassware, and linens are usually kept in a pantry or storeroom, and servers may have to retrieve these items before each meal. A steward is usually in charge of these items in large, formal operations. In some operations, these items might be located in the service station or on the service table near the dining area. This helps to speed service. Normally, servers get their own supplies for their stations, but buspersons may help them. Often, the server sets up the original stock and then, during the busy time of service, buspersons maintain supply levels and bring items to the station sidestand, or to the tables themselves.

Besides dishes and linens, salt and pepper shakers, condiments like mustard, ketchup, steak sauce, oil and vinegar in cruets, filled sugar bowls, and other items must be there, if their need is expected. All bottle necks on condiments, sugar bowls, salt and pepper shakers, and other items should be clean. Special attention should be paid to necks, mouths, and caps of bottles. **Exhibit 4.2** lists a suggested inventory of items for a sidestand.

Exhibit 4.2—Example of a Sidestand Inventory for a Station of 20 to 25 Covers

All quantities are recommendations. The items and quantities in individual operations will vary.

FLATWARE

Dinner forks	30	Soup spoons	24
Dessert/Salad forks	30	Cocktail forks (dinner only)	12
Dinner knives	30	Grapefruit spoons (breakfast only)	12
Teaspoons	30	Fish knives (lunch and dinner only)	12

(Other flatware according to menu needs)

DISHWARE

Dinner plates	25	Salad plates	25
Bread plates	25	Cups and saucers	25
Soup bowls	15	Bouillon or soup cups	15

(Platters and other dishware according to menu needs)

GLASSWARE

Water glasses or goblets	30
Red wine glasses	15
White wine glasses	15

(Other glassware as needed)

CONDIMENTS

Ketchup bottles	4	Dijon mustard	2
Worcestershire sauce	2	A-1 sauce	2
Tabasco sauce	2	Soy sauce (If needed)	2

(Chutney, oil and vinegar, grated cheese, etc. as needed)

REFRIGERATOR

Milk, regular	4	Milk 2%	4
Cream or substitutes, indiv.	36	Butter pats (144/box)	1 box

(Lemon, etc. as needed)

LINENS

Napkin, folded	30	Naperones (if used)	5
Tablecloths	5	Hand towels	5
Serviettes	5		

MISCELLANEOUS

Sugar substitute packets, salt substitutes, ice with tongs, syrups, jellies, jams, water pitchers, water, ash trays, pencils or pens, menus, peppermills, plate warmers, coffee makers, coffee packets, coffee maker cleaners, various trays, check trays, finger bowls, crumbers, tote boxes or buspans, etc., as needed.

Clean dishes can be brought to the sidestands on trays. The bottom of the tray should be lined with a clean cloth so dishes do not slide as they are carried. Be careful not to stack items too high; they may tip over and cause an accident or at least cause a server to lose time, and breakage can be costly. Cups and glasses can be brought to a service station in the racks they were washed in or on a tray. They should be placed inverted on the trays or in the racks.

Dishes, glassware, and other items that show cracks or chipping should be discarded, as should bent or otherwise damaged flatware.

Dishware, flatware, and glassware should be clean and sparkling. However, just because something looks clean and sparkling does not mean it is sanitary. Bacteria can be present without being seen. That is why these items are given a last minute or two of sanitizing while in the washing machines. In sink washing, the final rinse is extremely hot and usually includes a sanitizing agent. Wiping ware dry is not recommended. Health authorities frown on it because it can pass germs from one plate or utensil to others.

MACHINE CLEANING AND SANITIZING

High-temperature and chemical sanitizing dishwashing machines can help your operation handle a high volume of washing.

- Check the cleanliness of and clean each machine as often as needed, at least daily. Wash and rinse tanks should fill with clear water. Detergent trays and nozzles on spray arms should be clear.

- Flush, scrape, or soak items before washing. Pre-soak items with dried-on food.

- Correctly load the dishwasher racks—never overload them. This increases efficiency and helps ensure one-pass washing. Use racks constructed to expose all surfaces to the cleaning solution.

- Check temperatures.

- Check all items as they are removed. Run soiled dishes through again. Proper equipment and procedures will help ensure one-pass washing.

- Air dry all items. Do not use towels.

- Keep machines in good repair.

Hard water can cause spotting on items, and over time, a mineral buildup can occur. This spotting and mineral buildup can be avoided by using the proper detergents and other washing compounds, usually a compound associated with phosphoric acid. If spotting cannot be avoided, wiping might be necessary to remove the spots. Mineral buildup is not as easily taken care of. Soft abrasive material is needed to remove build up and it can be very time consuming. For these reasons, preventing the buildup in the first place is the best practice. Detergent companies often can be of help in solving build-up problems.

Sterling silver flatware and other items should be well polished. If the operation has a silver polisher, these items will come from the dishwashing section in satisfactory condition. If not, servers themselves may have to polish or touch up some pieces. Soiled flatware should be separated at the service station. Some operations allow flatware to be kept in the containers in which they are washed with the handles up. This is sanitary, but may not be very presentable.

In some types of service, additional equipment will be brought to the table. For French service, a *guéridon* with all the necessary items will be brought tableside so the *chef de rang* may prepare the items. Service carts or wagons (*voitures*) may be brought to the table to display items for guest selection, or a carving station or other mobile equipment may be brought to the table. They should be so clean they shine, and metal parts should be polished. Glass or plastic covers should be clear. All equipment—tray stands, high chairs, and other equipment that may be needed from time to time—should be clean, neat, and in good working order.

Servers must often operate equipment during the shift. It is important that a check be made to see if the equipment is able to hold items at the proper temperature.

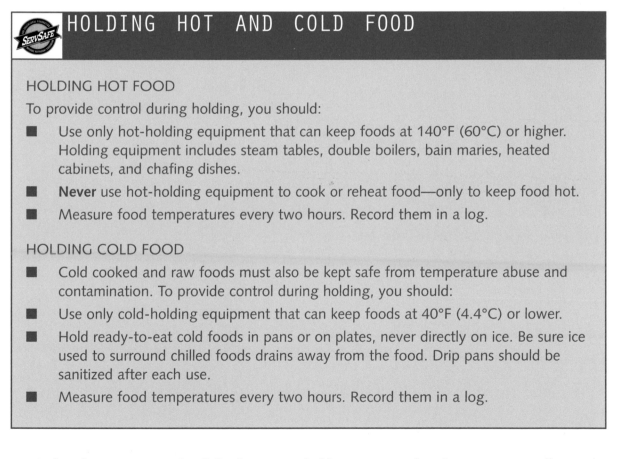

HOLDING HOT AND COLD FOOD

HOLDING HOT FOOD

To provide control during holding, you should:

■ Use only hot-holding equipment that can keep foods at 140°F (60°C) or higher. Holding equipment includes steam tables, double boilers, bain maries, heated cabinets, and chafing dishes.

■ **Never** use hot-holding equipment to cook or reheat food—only to keep food hot.

■ Measure food temperatures every two hours. Record them in a log.

HOLDING COLD FOOD

■ Cold cooked and raw foods must also be kept safe from temperature abuse and contamination. To provide control during holding, you should:

■ Use only cold-holding equipment that can keep foods at 40°F (4.4°C) or lower.

■ Hold ready-to-eat cold foods in pans or on plates, never directly on ice. Be sure ice used to surround chilled foods drains away from the food. Drip pans should be sanitized after each use.

■ Measure food temperatures every two hours. Record them in a log.

Before the station is ready, all food items needed by servers, such as butter, cream, coffee, ice, water, condiments, and garnishes should be on hand and properly stored. Usually such storage is close to the station and other servers may use it so the servicing of this area in mise en place is a joint responsibility. Buspersons might do this.

Frequently servers operate the coffee equipment. Before service starts, there should be a check to see that coffee and other items needed are on hand. Usually a 12-cup filter unit will be used. Coffee should always be fresh and prepared properly. Once a batch is made, remove the grounds immediately. If the grounds are allowed to stand over the brew, bitter compounds can drip down from them and reduce brew quality. Coffee should be held no longer than 45 minutes.

Table Mise en Place

The first thing a server in a full-service restaurant should check before placing linens on tables is the tables themselves. Are they sturdy and not wobbly? Is gum deposited under the edges? Are they clean? Are they level? Are chairs clean and the seats free of crumbs? Is the area around the table clean and in good repair? If not, any undesirable condition should be corrected before the table is set.

Some operations cover their tables with a **silencer** and tablecloth. The server brings the silencer and tablecloth to the table and places them on a tray stand or another table. If items are on the table, these are removed and placed on a tray stand, cart, or service station counter. The silencer is then placed on the table first; it might just fit the shape of the table top or it may hang down about eight inches over the side. To lay the silencer, pick it up at the centerfold at both ends and, with arms extended, lay it so the centerfold is in the center of the table and the silencer covers half the table on your side. Unfold the other half and place it so the other half of the table is covered. Sometimes the silencer may be given a quarterfold rather than a halffold. The procedure in this case is much the same as with the centerfold in the center, but the two top folds must be lifted up with the fingers and placed so as to cover the half of the table opposite the server. Now the underfolds must be carefully pulled from under and moved to the table edge where the server stands.

The method for laying the cloth on the silencer is similar to that used for laying the silencer. Smooth the silencer out before laying the cloth. The tablecloth should hang down at least eight inches over each of the table sides, with hangover equal on all sides. Some may wish to have the hangover so the tablecloth edge almost touches the chair seat. Smooth the tablecloth out and step back and check to see that the cloth is centered. If not, adjust the cloth. Examine all table coverings for holes, cigarette burns, and stains and replace them if necessary. A naperone, also called a laycloth, may be laid on the table in a similar manner to that used to lay the silencer and tablecloth.

The placement of napkins can vary. Some operations place them in a stemmed wine glass, although some health inspectors have raised concern because lint particles can remain in the stemware. Others place them folded in the center on the service plate, if used, or on the tablecloth between knives and spoons and forks, if no service plate is used. Place plainly folded napkins, either paper or cloth, with one corner facing the guest, so when the guest picks up the napkin, it can easily be unfolded with one hand. Often in fine dining establishments servers pick up the napkin and either hand it to the guest or place it across the guest's lap. Napkin folding has reached a high art of skill and many operations have intricately folded napkins that add to the table decor.

If the table is to be set with place mats or have no covering at all, wash the table top with a mild detergent in warm water and wipe dry. The tabletop should shine and have no sticky or soiled areas. Next, place the place mats squarely about two inches from the table edge. If these are the type that are reused, see they are clean and free from grime or stickiness.

Items that are to go at the table center, such as flowers, candles, and salt and pepper shakers should now be placed on the table, as well as ash trays, matches, table tents, and any other items. Then, proceed to set the table with the necessary covers. It is accepted practice to place flatware in the following manner: forks on the left, and knives (with blades facing inside) and spoons on the right. A dessert spoon or fork may be placed in the cover center above the napkin. Normally, flatware is set in place with the first piece to be used on the outer edge, and then as the meal and need for items proceed, the guest uses items from the outer edge inward.

The bottom ends of knives and forks should be placed at least one inch away from the edge of the table. When placing silverware on the tables, use a plate lined with a napkin to hold the ware. In casual dining, servers may use a cocktail tray, a plate, or simply a clean service towel as an underliner; servers should avoid walking around the dining room holding flatware with bare hands. Touching flatware by any part other than the handles should be absolutely avoided.

If the table set up requires it, the bread and butter plate is placed next to, but slightly above, the salad or appetizer fork. A butter knife is placed across the plate horizontally. Some operations prefer placing the butter knife vertically, so the tip of the knife points in the same direction as the dinner knife. In this case, the blades of both knives should face left.

The water glass is placed near the tip of the dinner knife. The wine glass (if required) is placed to the right of the water glass. If the operation offers two different shaped wine glasses (one for white wine, and one for red wine) the smaller of the two (usually the white wine glass) is placed slightly above the water glass, and the taller glass is placed slightly above and between the two. This might seem unimportant at first, but it will make a difference when serving wine as it is difficult to pour any liquid into a glass when a taller glass is in the way.

If additional glasses are used, such as those for sherry, champagne, and brandy, they should be set toward the center of the table, so that they will not cramp the space available between guests. This should be kept in mind when setting a table for four or more.

Whether in casual or fine-dining operations, symmetry is a must. An effective method to ensure symmetry in flatware placement is to draw an imaginary line between the tip of the dinner knife and the dinner fork on the opposite side of the table. If the line is straight, the setup is symmetrical. This can be applied to any table shape and size, with the only exception being a round table where an odd number of covers are placed.

Center pieces and condiments should not be placed in a manner inconvenient to guests; i.e., flower vases should not be in the way of conversation. Additional items that might be requested during the meal should be placed conveniently to the right side of the guest.

In some operations, such as a cafe, the table setup is simpler. Often there is no linen on the table and paper napkins may be used. A sugar bowl, salt and pepper shakers, ashtray, and matches (in the smoking section), and perhaps a bud vase or table decoration and some condiments are placed in the center. If the table is set against the wall, these items are placed on the edge of the table near the wall. Paper napkins may be in a container. Sometimes the table is set after the orders are placed, the server then bringing the correct items such as paper place mats, or flatware settings wrapped in paper napkins.

Often in coffee shops, cafes, or other busy operations, coffee cups might be on the table for breakfast. The flatware should be limited usually to a knife, fork, and teaspoon. Other flatware will be brought as needed, such as a steak knife, soup spoon, iced tea spoon, or dessert spoon or fork. Tables are often not covered with linen for breakfast or lunch—place mats will be used—while linen might or might not be used for dinner.

In most quick-service and self-service operations, the table is bare and guests bring to the table what they need. However, this can vary; some may still set tables with a limited amount of items.

Exhibit 4.3 contains a list of some basic mise en place instructions which will assist in any service style setup. **Exhibit 4.4** is a reminder list of common side work.

Exhibit 4.3—A Partial List of Mise en Place Instructions

1. Be on shift and station on time.

2. Be sure to punch in on the time clock when you arrive.

3. Be prepared regarding personal appearance and uniform.

4. Check your station for adequate supplies of glassware, garnishes, straws, napkins, picks, etc.

5. See that all equipment is clean and ready to use.

6. Sign in for and receive change money.

7. Check station to make sure all areas are in proper condition (tables/chairs clean, arranged properly, floor clear, ashtrays on table, etc.)

8. If you are relieving another server, check with her/him for any instructions as to guests that might be in the station, pending orders, etc.

9. The service area at the bar is to be kept clean at all times, free of litter, soiled towels, dirty ashtrays, etc. All garnishes and supplies must be stored in an orderly fashion. Wipe the bar service area frequently to keep it clean. Keep garnish containers clean and filled with fresh items. It is advisable that different servers on each shift be made responsible for specific service area tasks; thus, one server can be made responsible for garnishes, another for picks, napkins, etc.

10. Be sure you have a bottle and can opener at your station.

11. Be sure the ice bin area is kept clean. Use the scoop whenever filling glasses with ice. Every station should keep about a dozen glasses iced up.

12. Drink tower areas and sinks for drawing fresh water should be kept clean. Do not wash ashtrays under drink towers. This may create serious drainage problems.

13. Always use a large service tray in giving service. Prepare and stack at least six trays. Arrange service items on trays on one side of cocktail tray. Each tray should contain:
    - ashtrays
    - tip trays
    - matchbooks
    - pens
    - bar towel
    - cocktail napkins

14. Check ashtrays frequently. Remove as necessary, replacing with clean ones. Use the capping method by placing the clean ashtray on top of the dirty one; remove from table and place the dirty ashtray on side tray. Then put clean ashtray back on table.

15. The floor around your station must be kept free of litter (straws, napkins, empty cigarette packages, etc.)

continues

16. Unoccupied chairs should be pushed close to the tables to allow free movement between tables. This will also help to keep the station organized.

17. At the end of the shift and at closing, all tables and chairs should be returned to their proper places as per station chart. They should all be wiped clean with a bar towel.

18. At the end of a shift, all ashtrays should be washed and returned to their storage area in the station.

19. S-O-S (stay on station). Be ready to offer more service. Take that second order. Pick up soiled napkins, discarded stirrers, etc. Put chairs back in place.

20. Circulate; don't congregate.

21. Utilize the round trip. Don't return to the bar empty handed. Clear tables before going to the bar. With light loads from the bar, bring station supply replenishments.

Exhibit 4.4—Side Work Reminder

A. Condiments
   1. Fill partially full containers with other partially filled ones. Fill to half-inch from the top.
   2. If containers are to be refilled, wash first and then refill.
   3. With a damp cloth, wipe tops of containers; clean caps and replace.
   4. Store perishable condiments in neat rows in the refrigerator.
   5. Day servers will empty and refill relishes.

B. Sugars, salts, and peppers
   1. Fill and wipe salt and pepper shakers; make sure holes are clear.
   2. Refill sugar trays. Be sure old packets are not allowed to rest on the bottom under the row of packets on the tray.
   3. Wipe holders.
   4. On Monday nights all containers and trays should be emptied and washed. Refill only after being sure the containers are dry.

C. Butter and cream
   1. Don't use broken or melted butter patties. Place broken pieces in dish to be taken to kitchen for use in cooking.
   2. Keep butter containers iced and free from melted water.
   3. Return emptied butter containers to pantry for washing and refilling.
   4. Keep coolers clean at all times; wipe out as needed. On Sundays, empty and wash, refilling with butter and cream.
   5. Fill creamers to about 1/2 inch from the top. Do not refill creamers; use only clean ones.

continues

6. At end of shift, empty filled creamers into cream containers and send the emptied creamers to the dishwashing area.

7. Keep at least two containers of cream and two filled butter containers in the cooler; more may be needed at busy times. Pick these up at the pantry.

D. Dining room buffet and water cooler

1. Have place mats, napkins, salt packets, sugar packets, and takeout containers on hand.

2. Have clean water pitchers on hand.

3. Stock all these items on the shelves under the buffet.

4. Stock tip trays and ashtrays on top of buffet. Keep the remainder of this area free for use during service.

5. Stock glasses under water cooler.

6. Remove daily the tray under spigot and wash drain area. Once a week use drain cleaning solution. Wipe down entire cooler daily.

E. Coffee urn area

1. Clean iced tea dispenser and tray daily.

2. Empty ice cubes for closing work; return lemon slices to kitchen pantry.

3. Stock ramekins, iced tea glasses and spoons, tea pots, children's cups, cups, and saucers on undershelves of coffee urn stand.

4. Stock coffee, tea bags, instant tea bags, and coffee filters on undershelf of urn stand.

5. Wipe entire area down daily.

6. Clean coffee urn thoroughly at the end of each emptying. On closing, leave urn filled with water that has had a fourth cup of soda added. Mix well after adding. Rinse urn thoroughly in the morning after emptying and before using. On Saturdays add one bag of cleaning solution instead of soda and mix well. (Store this cleaning solution in box on bottom shelf.)

7. Send bar trays through dishwasher once a week.

F. Bread warmer, boards, and knives

1. Wipe bread boards and bread knives with damp cloth after each use.

2. On closing, empty all trays of bread supplies and take to bake shop. Clean drawers. Be sure they are free of crumbs.

3. Wipe entire outside of warmer daily.

4. Have a good supply of napkins.

5. Clean bread baskets once a week.

G. Salad dressings

1. On closing, empty dressing jars into their storage containers and place the storage containers back into the refrigerator.

2. Stack the empty containers on a tray with ladles in between and take to dishwashing section.

continues

3. Wipe racks with a damp cloth daily. Empty and clean refrigerator once a week.

4. Store condiments on table top. Clean tops and store perishable ones in refrigerator on closing.

5. Store French and Italian dressings on shelf under shrimp cocktail containers.

Guest Checks

Order or guest check systems vary from operation to operation. Some restaurants issue servers a supply of guest checks; with computerized POS (point-of-sale) systems, this usually is unnecessary. Instead, checks, properly coded as to server, station, and table, are given to the server at the time guests are brought to the table, or are produced once order information is entered into the computer.

Where paper checks are used, servers are usually given a set quantity of numbered checks. Servers sign for them and are responsible to see that guest checks are properly used and that they get to cashiers with payment. In some operations, checks are issued to servers at the time guests are seated. These checks are kept at the cashier stand. Servers sign for the checks given to them at the start of the shift, usually next to the check number on a form commonly called the **traffic sheet**.

GUEST CHECKS

Cashiers, servers, and bartenders may be able to steal money from the establishment by manipulating guest checks. Watch for a slower flow of checks from the bar or dining areas, missing checks from the check supply, fake guest checks among the legitimate ones, or checks that are regularly turned in late by an employee. Techniques used by employees may include:

- Destroying checks, not ringing the sales, and keeping the money.

- "Bunching" checks by ringing only one of several checks with identical amounts, collecting all the money, and keeping all but the money for the one rung check.

- Giving incorrect change to guests.

- Raising prices on checks or charging for items not ordered.

- Giving the guest one check, ringing in a smaller check, and keeping the difference. Checks stolen from the establishment's supply, duplicated, or brought in by the employee may be used to do this.

- Receiving payment from another server's table, keeping the money, and replacing the check on the deserted table so the other server thinks the party has "dined and dashed."

GUEST CHECKS (CONTINUED)

A check is made to see that all server checks are accounted for. This helps to see that guests do not skip out without paying. A few operations require servers to pay a certain amount if a check is missing. POS systems trace a check and alert managers to discrepancies. If guests are given the checks to be taken to the cashiers for payment, servers are not usually considered responsible. Cashiers then need to stop guests and get payment. Some guests wait until a group is at the cashier's station and then slip out. All employees should watch to see that this does not occur.

CUSTOMER FRAUD

Diners may purposely fail to point out when a bar tab is not carried over to their food order. They may also alter or erase entries on guest checks left on their table. Careful register use, stapling all related guest checks together, and keeping control of the checks as long as possible can help avoid these schemes. Customers may also claim they were given insufficient change. To avoid this complaint, the register operator should leave the payment on the register until the change has been counted out to the customer, and then put the payment in the drawer.

Getting Servers Ready

Scheduling

In full-service restaurants, specific mise en place planning must be done by the host, hostess, maitre d', or other head of service. This includes forecasting server staffing needs. This is typically done by estimating the total number of covers, and then dividing this total by the number of covers one server can take care of during the meal period. Thus, if an operation estimates 120 guests for lunch, and the number of covers a server is expected to serve is 40, the number of servers scheduled is determined: $120 \div 40 = 3$. The forecast may be made for a week, and daily work assignments should be made so servers know when they are to be at work. Servers should be informed of their schedules well in advance so they can plan their other priorities.

It is important to plan the number of servers needed and the station assignments on a practical and realistic basis. Each operation's staffing needs differ based on the size and type of the establishment. Only the number of servers needed should be scheduled, and no more. If more customers come than were expected, or a server misses work, a well-organized staff of servers can smooth over the unexpected and still have happy, satisfied guests. It is important in such cases that

the service staff be well-trained to ensure that things run smoothly.

Wisely, some operations maintain a list of "extra board" service employees, which are not included in the payroll on a steady basis but are called in case of emergency. Larger operations use an "on call" system.

The person in charge of service will assign stations. Station assignments let the servers know which stations they will be working, so they can move ahead and complete the necessary mise en place. Often a chart is posted showing station locations and the name of the server(s) responsible for them. As the customer flow fluctuates, stations may be restructured or reassigned. This allows for servers to cover stations while others take a break or end their shifts.

The Pre-Shift Meeting

Operations often have a short meeting of servers or all employees—sometimes called lineup, or pre-shift meeting—just before the meal period begins. Tables should be set, all other arrangements should be completed, and only last-minute things should be done. The meeting is usually conducted by the manager, host or hostess, or a supervisor. Often a quick inspection of grooming and dress occurs, but this is informal.

This pre-shift meeting is to go over the menu and any details of service that need to be brought to the attention of the service staff—whether any menu items are not available that day, the soups or fish of the day, etc. Perhaps some important guests have made reservations or a special group is coming in which should be brought to the attention of servers in the station where they will be seated. Any special menu items should be covered with a description of how they are prepared and served. The correct items to use with various dishes and the manner of the service of the items, along with price and preparation time, should be included.

Each server should copy this information down since it may not appear on the menu and all information will have to be given verbally. Specials are usually items management wishes to promote, but others on the menu may be highlighted. Planned runout times may be given and suggested substitutes indicated. A bulletin board in the kitchen can give some of this information. A visit may be made to the kitchen where the chef may briefly add any details he wishes the server group to hear.

Sometimes the meeting covers complaints or compliments received. These can be helpful in indicating trouble spots and also indicating what guests like about the service or food. Since the time that follows can be stressful, the meeting should end on a positive note so servers are motivated to deliver great service.

Clearing and Resetting Tables

When a full-service dining room is full and other guests are waiting anxiously to be seated, tables cannot remain vacant long without angering customers. Clearing tables, changing table linen, and resetting must occur rapidly. Servers, buspersons, host(s) or hostess(es), and managers must work together in this regard.

Changing the linen is more difficult than in the original setting up. Soiled linen, napkins, flatware, table decorations, salt and pepper, and other items must be removed and placed on a traystand or sidestand—never on a chair or another table. If the table is against the wall, it is pulled a few inches away for clearing. To prevent crumbs from falling from the soiled cover onto the chair seats or the floor, fold the soiled cloth in the center and then bring the ends together, sealing in the crumbs. The soiled cloth should then be placed on the traystand or sidestand. The server is then ready to place a fresh tablecloth on the table. The table is reset with the proper dishes, flatware, and glassware, and items on the traystand are put back on the table. Napkins are then placed accordingly.

When a bare table is used, the table is wiped clean with a cloth dipped in warm water with detergent and wrung completely dry. Once the table is re-set it is ready for your next guests.

Ending the Meal

Mise en place is also required to prepare for guest departure. The bill should be totaled and ready to deliver. Some facilities deliver the bill by placing it face down on a small tray in front of the person paying, or in the middle of the table if that is not known.

If the payment is by credit card, provide a pen for writing down the required information. A rapid return with the change or the credit card and credit charge slip should be made. Any personal items that servers have taken care of during the meal, such as crutches, should be quickly brought. Help may be needed to put on coats. A sincere thanks from the server is appreciated. As the guests leave the dining area, the host, hostess, maitre d', or other individual at the entrance thanks them and invites them to return. In very informal situations, such as a quick-service operation, this may not occur. Payment is made when the guest gets the food, and there is little formality needed on departure. Nevertheless, when guests are leaving, a warm thank-you makes all the difference.

The departure of guests should be given as much attention and care as the arrival. Cashiers, coatroom attendants, managers, doorpersons, valet parkers, and others should be friendly and helpful.

As a shift winds down, servers should complete their sidework. All soiled dishes should be taken to the dishwashing area and clean stock should be replenished. Soiled linen should be taken to its designated receptacle. Salt and pepper shakers should be filled, cleaned, and checked to see that holes are not plugged. Sugar bowls should be filled and cleaned. All other condiment containers should be filled and cleaned according to the operation's standards. Periodically all salt and pepper shakers, sugar bowls, and other equipment should be emptied, washed thoroughly, dried, and then refilled. Check the sidestand to see that it is clean and the inventory is up to par. Old coffee should be emptied out and the containers cleaned. It might be necessary to put into the coffee machine a compound that frees it of buildup on a regular basis.

Health regulations in communities differ and must be followed. Perishable condiments should be placed in a refrigerator. Iced or refrigerated butter may be returned to refrigeration. Items not in sealed containers that have been put onto guest tables should never be reused for service. However, items such as jams or jellies in sealed containers are reusable. Be sure to know the expiration date for all perishable products and discard any items that have not been kept sanitary and at a safe temperature.

The closing work differs according to the operation and the meal. In most operations, closing includes setting up the dining and service areas for opening the next day. Each operation should establish complete checklists that outline what must be done at the end of each meal period to prepare for the next. For maximum efficiency, tasks may be assigned to individuals based on service area, shift length, etc.

The inspection checklist in **Exhibit 4.5** can be used at closing, as well as before guest arrival. Servers should leave their stations clean, stocked, and prepared for the next shift. The next server will appreciate help on getting a good start. Likewise, each server profits from the servers who previously worked the station.

Exhibit 4.5—Sample Closing Checklist

Inspected by_____

Time and Date _____

Yes	No	
❏	❏	Lights off except designated ones?*
❏	❏	Air conditioning turned off?
❏	❏	Proper doors, windows, etc. locked?
❏	❏	Alarm set properly?
❏	❏	Tables cleared properly?
❏	❏	Chairs inverted on table tops?
❏	❏	Perishable food supplies cared for?
❏	❏	Cups, glasses, etc., inverted on trays?
❏	❏	Sidestand top cleared except for allowed items?
❏	❏	Heating and other equipment turned off?
❏	❏	Soiled linen taken to soiled linen receptacle?
❏	❏	Traystands properly stored?
❏	❏	Proper amount of flatware, dishware, and glassware on sidestand for breakfast?

*Leave lights on for the inspection; turn off just before leaving.

Quick-service Mise En Place

Attention to detail is as important in quick-service operations as it is in full-service restaurants. The primary servers are the counter employees, and employees out on the floor cleaning and stocking amenities. When taking the order, the server should repeat the order so the guest can hear and verify it before it is totaled up on the preset register. Payment is usually taken before the food is served.

When the order is assembled and put on a tray or in a bag, the guest typically takes it to a table or a service area for eating utensils, napkins, and condiments. In some operations, guests bus their own tables and dump their own trash. In others, employees perform this task.

In operations with drive-through pickup windows, the employee is responsible for helping guests select items, taking and placing orders, expediting, and giving a courteous, positive impression to guests.

To give such service requires precise and considerable mise en place. At peak service times, long lines challenge servers to work quickly and accurately. Without good mise en place, quick service is doomed. For service staff, this means that plenty of bags, wrappers or containers, condiments, coffee stirrers, cream and sugar, napkins, and straws must be on hand. Cash drawers should never be allowed to get low. Only the very last-minute details of assembly should be left. Because of the speed and efficiency expected by guests, mise en place is crucial.

The Cash Bank

In some operations, cocktail and other servers carry their own **cash bank**. This allows quick payment of a check and reduces the amount of work required of the staff. Procedures for some servers may have to make cash transactions from a cash fund. If used, one of the first steps of coming on shift is to obtain a cash bank to use in making change when collecting bills. The server should be sure to verify the correctness of the sum given.

Servers may carry the bank, but often the bank may be kept at a cash register in a drawer that only that server can operate. The register must be turned on and the server must enter his or her server number, perhaps a password, drawer number, etc. The check for payment will be inserted into the machine allowing it to read information on it, such as check number, table number, server number, items on the check with their price, and check total. The amount of payment is entered; the proper sum is put into the register and the drawer is closed, which closes the transaction. The check is filed in a special spot or may be kept by the server for closing out the cash bank at the end of the shift.

Sometimes cash is not given by the guest but some other type of payment is used. A common way of paying in clubs and hotels is to charge to one's account. In clubs, no identification is usually needed. If one is needed, the club member usually makes identification by offering a club membership card. In hotels, the person charging the bill may show for identification a key or key card or some card issued at the time of registration. If there is any doubt as to identification, a call can be made to a front desk or other agency to verify a name, room number, etc.

Credit card use is common. Servers must know what credit cards are honored by the operation. Some operations subscribe to some system of credit card verification which indicates whether or not the card should be honored. The server should use this system and, if the card is verified, make out the charge slip, giving the card number, total, etc. The card with the slip is brought to the card owner who signs the slip, puts in the amount of tip and receives a slip copy along with the credit card and a sincere thanks. On closing out the server's bank, these slips are offered instead of cash. Some operations ask the server to fill in a sheet indicating credit card charges and total these. A sheet reporting complimentaries, discounts, or other such transactions may also be required in the closing report.

At times, corrections must be made on ring-ups. An over-ring may be corrected by processing a void system set up by the operation. Similarly, a discount or complimentary system may be set up which servers can use to indicate a lower payment than indicated on the check. Often, some type of ticket, card, or other item is offered by the guest to verify the right to the lower payment and this is filed with the checks. If there is an over-ring, the server may have to write this on the check and also fill in the transaction on an over-ring sheet.

When no cash register is used, and the server carries the bank, discount and complimentary records are usually written up by the server on some record sheet. Servers must have an adequate mathematical and writing background to complete such reporting.

At the end of the shift, the register may give the server a total printout of transactions made by that server during the shift. The operation may also wish to have a record filled out recording the check number and check amount with a total sales given at the bottom. Servers must accompany such a sheet with the proper amount of money. When servers keep tips in the register, any overage in the register after the amount due on sales is set aside and belongs to the server. The server presents the original cash bank, the sale money, and other records to the appropriate authority for a signed release, and is freed of all further responsibility unless some discrepancy is found in the report or funds turned in.

Chapter Summary

Mise en place—preparing ahead for the service period—is essential if excellent service is to occur. Time is saved and serving is organized. Pre-preparation of needs should be planned and accomplished to cover all phases of a guest's visit, from the time of arrival to departure.

Grounds must be neat and clean. Good security and safety are important. Entrances, hallways, phone booths, coatrooms, and other areas guests encounter upon arriving require attention. Sanitation standards also are a must for today's service personnel.

Mise en place work needed to care for guests can be divided into three parts: 1) getting ready, 2) between-guests work, and 3) ending the meal. The first covers all preparations needed for the arrival of guests until they arrive at the dining area, and the work that has to be done before a meal starts, such as getting all items and condiments ready, getting checks, having a lineup meeting, etc. The second part covers the activities that take place when one group of guests leave and the table must be readied for another. It requires fast work if the dining area is busy and guests are waiting for tables. The last part consists of the preparations that must be made for guest departure and the work that must be done to see that the dining area is ready for the next meal.

Chapter Review

1. View several different kinds of facilities and look over the grounds, the entrance, and the way into the dining area. What do you find good and bad? What would you do differently?

2. What are some of the requirements of valet parkers?

3. How can a busperson assist the server in mise en place duties?

4. You are a server and have just arrived at the station assigned to you. What tasks should you perform, and in what order should you perform them?

5. What is a pre-shift meeting? What is its purpose?

6. How do some computers help stop security problems?

7. What is the proper way to clear a table of items before resetting it?

8. When guests are preparing to leave, what mise en place must be done to see that their departure is facilitated and they receive adequate attention to make them feel they were welcome?

9. Why shouldn't you wipe dry wet dishes and glasses?

10. How does one forecast the staffing needs of an operation?

Service in Various Industry Segments

Outline

Key Terms

Wave system

Flying service

Flying platter

Shopping service

Sneeze guard

Sweet table

Action station

Aboyeur

Learning Objectives

After reading this chapter, you should be able to:

- Describe proper meal service and clearing for banquets, specific meals, buffets, and other types of service.

Introduction

The foodservice industry is one of the most varied in the world. This chapter offers an overview of how service is handled in many segments, from buffets to lunch counters to hotel room service.

Banquet Service

Photo courtesy of the Chicago Hilton and Towers.

Regular Banquet Service

A banquet, whether a breakfast, lunch, or dinner, can be a special celebration, a gathering to honor someone, or a professional meeting, often preceded by a reception with beverages and hors d'oeuvres.

Many details go into managing a banquet. Some customers request dancing or entertainment between courses. This means there must be a way to provide music and a stage or floor for performing or dancing. Special flower arrangements may be requested. If there is a speaker or some type of program, the arrangements for this need to be agreed upon beforehand so everything can be ready. A sound system, photographers, special lighting, and a host of other items may also be needed.

The menu must be discussed and approved in advance, along with the total cost. The service style may be specified, or the facility may plan its own. At least one week before the banquet, the number of servers needed must be decided and an appropriate schedule should be created. Banquets usually require a special service staff. This must be taken into consideration when creating the service schedule. All servers should be aware of the proper dress or uniform for the event.

American and Russian service are the most common service styles used. In American service, food is placed onto the plates from which it will be eaten and brought to guests. In Russian service, all cooking, finishing, and carving are done in the kitchen. Only dishing the food is done at tableside. Sometimes, a separate crew sets up the room with the bare tables and chairs. Some operations have their own employees lay the table linen and set the tables. Set-up time can easily be worked into the service schedule.

A sample setting is usually provided so those setting up the room can refer to the model to follow. Often the setting up is a team effort with some laying linen, others setting silverware and glassware, while others set up salt, pepper, and other table appointments. Dishes and other items for the affair may be in a storeroom pantry. It is helpful and efficient if these items are on mobile carts so they can be rolled out to the dining area.

In standard functions, all flatware, china, and glassware are present while in more elaborate functions, the initial setup consists simply of a knife, a fork, and the water glass. In this case, before a course or wine is served, the proper silverware and glassware will be placed on the table. This may require a larger number of service staff and will increase the price for the event. When pre-setting all tableware there are two definite disadvantages: A large amount of tableware cramps the space, making it very uncomfortable for the guest, restricting freedom of movement and social interaction, and, when first seated, guests are often intimidated by seeing ten to fifteen pieces of flatware and countless glassware items in front of them.

In some banquet setups, coffee cups are preset. They should be positioned to the right of the guest but further up along the same line of the glassware so as not to cramp the space between the guests. The handle should always be at a 3 o'clock position. Generally, each cover needs at least 25 inches of linear space; if the cup is placed downward it will be too close to the bread and butter plate of the guest to the left.

Some managers do not feel that any food items, except ice water, should be on the table until guests are seated. Others may wish to have the first course, if cold, on the tables when guests enter. This may be done to speed service and may be done at the request of those in charge of the banquet. However, some operations even go farther. Not only will the first course be on the table but the bread and butter will be also.

A lineup or pre-shift meeting may also be used to give specific instructions on setting up. If the servers are accustomed to working banquets at the facility, the lineup meeting may be simple with just the special needs discussed. If servers are new, then the meeting must cover procedures in detail with perhaps some repetition and emphasis to see that special needs are not overlooked.

Tables will be arranged as needed for the banquet according to a pre-arranged floor plan. Often there is a head table facing into the room. The speaker podium and other items will be set up at this table. Often this head table is on a raised platform, or dais, so it can be more easily seen. Various additional table arrangements may be used, but the most common is a 6- to 10-cover round table. Tables should be numbered, not only to let guests know where their table is, but also for assigning stations. Before guests enter the banquet room, it is important that the service areas are clean and presentable. A detailed check should be made to see that the arrangement of settings and tables is symmetrical and attractive. An individual in charge of service, and perhaps some assistants, are usually on the floor during the service to see that the function proceeds as planned and that every detail as specified in the banquet function sheet is handled.

Station assignments may be by individual server areas or by teams. When the **wave system** is adopted the entire banquet room becomes one large station. Once a starting point is established, the tables are served in sequence by all servers according to a pre-established order. Experienced banquet service personnel who have been working together for some time often prefer this method. Service can be synchronized so that while some servers clear the table after the various courses, others bring the next dish and serve it. It is customary to start service with the head table and to quickly follow this with service to the other tables so all guests start and finish eating at about the same time.

The wave system offers some advantages, but in a typical banquet situation, the individual station method is still preferred. A normal station for one server may include from 24 to 32 covers. This may be increased or decreased according to the complexity of the menu. The station that includes the head table may have fewer covers. In some areas the local union contract establishes the maximum number of guests to be served per server.

On the average, the dining space required is 15 to 18 square feet per guest. The dining room layout depends on many different factors such as the type of event, the menu, the number of guests, the room shape, and the host's preference. Round tables are preferred in banquets as they are more suitable for conversation and more practical for service than any other shape. Sizes may vary although the majority are either 60 inches in diameter (to accommodate six guests) or 72 inches (to accommodate eight to ten guests). Rectangular six, eight, and ten footers are also common. Square, oval, half round, and serpentine-shaped tables are used less frequently. Unless otherwise requested by the host, round tables are placed four to five feet apart to allow sufficient space for chairs and service traffic. Rectangular and other shapes would require additional space, normally five to six feet, although with the same space availability a larger number of patrons can be seated by using rectangular tables as opposed to round ones.

For each table size and shape there is a corresponding table cloth, therefore a 60-inch round table will require an 85-inch-by-85-inch cloth while an 8-foot rectangular table (96-inch-by-36-inch) will be properly covered with a 54-inch-by-120-inch cloth. Special occasions and specific needs may require the various shaped tables to be placed in configurations such as a "T," an "L," a "U," a "horseshoe," a "comb," or "E," an "imperial table," or a "block," just to name a few.

Since everyone eats the same meal at the same time, the service is simplified. This is helpful because it is essential that fast service occur. Usually banquets are planned on a tight schedule; the meal must be over quickly so that the program to follow is not delayed.

Removing dishes after the program has started, or otherwise having server activity in the dining area when the program is going on, is very distracting and can lead to guest dissatisfaction with the whole function. Only if requested by the banquet host should the servers be allowed to walk around the tables or perform service duties during a speech, an award presentation, or any guest activities. In a typical situation, if a server is placing a tray of hot food on the jack, or stand, ready to serve, when a guest initiates an unscheduled toast or blessing, the server should wait for the guest to finish before continuing service.

Servers should learn the menu so they can plan their work, give smooth, efficient service, and provide competent answers to guests' inquiries about the items served. It would be a disappointing situation for the guest and an embarrassing one for the server if an answer could not be given in regards to the type of wine to be served, why it was selected to match various courses, what beverages are available, etc. During the meal various other items may be needed. For coffee service at the end of the meal, cream and sugar may have to be brought to the table. Planning ahead and preparing these items is helpful. A complete mise en place, as discussed in Chapter 4, is essential for a satisfactory banquet service flow.

In some operations, the food items are dished up in advance and the plates are covered and placed over sheet pans on heated sheet-racks on wheels, or similar holding containers. This method makes serving food a much simpler task for servers. Once the racks have been wheeled out to the dining room, the plates can be readily picked up and served. However, this practice may not be recommended if the food being served does not hold its appearance well. For example, after being in the various holding cabinets or sheet-racks for more than three minutes, sauces may develop a "skin" and garnishes can become soggy.

Dessert service may be spectacular or something simple. A special occasion cake may be cut ceremoniously, or simple sherbet or ice cream may be placed on the table. Depending on the specifics of the event, there can be a wide range of options for serving desserts.

If wine or any other special beverage is to be served, often special pourers will perform this service so the regular servers can focus on serving the food. Whoever does the pouring should know the essentials of such service so it is done correctly. (Such service is discussed in the **Bar and Beverage Service Module**.) However, some steps in traditional wine service are eliminated. The wine bottle is not shown to the host before pulling the cork, the cork is not pulled at the table, and no one is asked first to taste the wine to see if it is suitable for the occasion. These steps are unnecessary, as the host has usually specified and approved the wine selection prior to the event.

Those hosting the banquet must provide an estimate of the number of expected covers. A facility might allow a 10 percent overcount or undercount on such an estimate. Whenever guests must present a ticket or some other entrance permit, this may be used to confirm the count. Otherwise, a head count may be taken, but this can be difficult if the number of guests is large, such as over 2,000. Another way to confirm the number of covers is to have a plate count. The number of plates not used subtracted from the total plates set out may be taken as the count. In this case, those hosting and paying for banquet must accept the word of those making the plate count. Another way is to count the number of full tables and multiply by seats per table, then add the counts of the partial tables.

Banquet Table Service

As previously discussed, the two predominant service styles in banquets are American and Russian, with the American being the most popular. When serving courses according to the basic American service style rules, all foods are served from the left side with the left hand and cleared from the right with the right hand. Plates should be placed gently and attention must be paid so that they are positioned at least one inch from the edge of the table. All beverages are served from the right and cleared from the right. Left handed servers might make some exceptions in clearing the plates if they need to handle dishware with a firm grip.

Once a tray of food is placed in the service stations, some servers like to pick up four or five plates at once as they feel they can service the whole station faster. It is recommended that no more than three plates be carried at one time unless they are small and "dry." When plates are ten inches in diameter or more and they contain thin sauces, the risk of spillage increases. Carrying a maximum of three plates is also more aesthetically acceptable.

When pouring coffee, banquet servers often lift the saucer and cup from the table as they feel it is safer and less difficult. In reality the opposite is true. Coffee cups or any beverage container should not be removed from the table. To pour over it while the cup is resting on the table is actually safer as well as technically correct.

Breakage is a main concern of banquet operators, particularly during large functions. It is a known fact that banquet rooms have more breakage than any other type of foodservice operation. When collecting soiled dishes, items should be carefully separated and plates should be stacked according to size. This not only will minimize breakage but also speed up the breakdown procedures and will be of great help to the dishwashers. Banquet servers normally carry trays on the left hand, flat, resting their weight on their shoulder. More experienced servers may carry heavy trays on the tips of the fingers. This should not be attempted by a novice as it could result in breakage, embarrassment for the staff, and injury to a guest. (Refer to Chapter 7 for tray carrying guidelines.)

It is imperative that servers fully understand how crucial timing is in banquets. In the lineup meeting discussed earlier, the sequence of service should be addressed in great detail. Without proper planning, a smaller unforeseen incident might throw off the timing with disastrous consequences. The wise server always includes a minute or two for something unexpected in the sequence of service time flow. For example, in a banquet of three hundred guests where a prime rib dinner is served, guests usually will not question the cooking temperature of the meat. If a guest believes that their cut is too rare and asks that the meat be cooked a few more minutes, their request should be honored. One dish alone can disorganize an entire station as the server has to make special accommodations for the guest. Meanwhile the station is unattended and patrons are complaining. This is the time for other servers to step in and help. A fellow server can help keep an eye on the "abandoned" station. In some operations, servers play it by ear and proceed according to the individual station needs at any particular moment. Teamwork is essential to successful banquet service.

Buffet service should be used only with fairly small banquets. A large number makes for too much milling around and a lot of confusion. It is also difficult to move along quickly. If buffet service is used, the number of courses may be limited and often servers clear the table after the main course, then serve the dessert and beverage courses.

Receptions

Receptions are gatherings to celebrate special occasions or to honor a person or persons. They may be small, as for a small group of friends who are invited to meet a bride and groom, or they may be a large reception preceding a banquet. Often there will be a service bar or table at which beverages will be offered. Beverages may include: punch, champagne, wine, beer, cocktails, and soft drinks. Nonalcohol beverages are becoming more common on reception menus.

If guests are expected to carry their beverages and hors d'oeuvres, servers will be needed to take empty plates, soiled serviettes, and beverage glasses from them. Tray stands with trays should be conveniently located away from sight, if possible, where these items may be placed before being returned to the kitchen for washing and sanitizing. However, instead of a bar or service table, servers might present trays of beverages and hors d'oeuvres. This is called **flying service** and the trays carried by servers are called **flying platters**. The Russians influenced the original French reception custom and introduced the concept of standing while eating.

Showroom Service

Showrooms often offer a very good meal and excellent entertainment at a nominal price. This attracts patrons to the facility, where gambling occurs. The showroom with its food, beverage, and entertainment costs may lose money, but the loss is made up by the gambling department of the company. Service very similar to that of a showroom may also occur at a race track or other entertainment facility. Again, the price may be nominal, the purpose being to bring in patrons who will gamble on the races or otherwise spend money.

Sometimes, only beverages are offered, and this is most common for the late night show or for cabarets. Both food and beverage service and the entertainment are included in one cover charge. The cost usually includes a preset menu. Tips and additional items are extra. Showroom menus often offer a fairly elaborate meal with an appetizer, a salad, a main course that may be steak, prime rib, poultry, fish or seafood, along with a dessert and beverage. Beverage alcohol is usually not included, although in some cases a pre-dinner cocktail or after dinner drink may be included in the cover charge.

Often a showroom facility has a large number of reservations and, when the door opens, these people are seated immediately. Seating 500 or more people within a few minutes is a challenge and requires good organization and planning. The service crew is usually experienced, working steadily at the operation for perhaps several years. If the operation is unionized, the union provides the required number of personnel. In one Las Vegas operation, as many as 35 teams of two servers may be used to serve about 1,000 covers. (Today, servers in casinos tend to only serve cocktails during shows.) American service was the most common service used in these situations.

After the orders are taken they are brought immediately to the kitchen. Some operations have an announcer call out the orders to prevent confusion. This also allows servers time to do other things. Serving moves along rapidly since there are only a few pre-set items and these are often pre-portioned and ready to be placed onto the serving dishes.

Setup procedures for showrooms are the same as for banquets, with groups of servers working together to place all the necessary items on the tables. The servers are also expected to see that the required settings for the following show are ready at the end of the previous show. As soon as the show ends, servers must enter the room and quickly lay clean linen and set the tables. There is a limited time before the next group of guests enters.

Kosher Meals

The presence of an expert in Jewish law is required during the preparation of kosher foods because it can be quite complex. Traditionally, kosher foods are served in festivities and catered events such as bar and bat mitzvahs and weddings.

All foodservice employees should be aware that the Jewish dietary law strictly forbids that meats, and all foods prepared with meat products, come in contact with dairy products. Kitchen or dining room equipment used to prepare meat products cannot be used prepare dairy products. Kitchens used for kosher banquets have separate sinks, equipment, and different sets of working utensils. This prevents the mixing of meats and dairy items in preparing, cooking, and serving.

Servers should become familiar with the principal characteristics of kosher foods. The term *kosher* means "fit" or "proper." All kosher foods must be handled, prepared, and served in full accordance with Jewish dietary law. This law allows only the consumption of meat from animals that have split hooves and chew their cud. Pigs have split hooves, but since they do not chew their cud, pork is forbidden. Common birds such as chicken and turkey are allowed to be eaten but not scavenger birds or birds that are hunted. Most fish are allowed, while shellfish are not.

Service for Specific Meals

Breakfast

Breakfast service is usually faster than other meals because guests are often in a hurry, except at group gatherings where the occasion may be one where time is not so vital. Sometimes breakfasts are numbered on the menu so guests can pick out a group of foods and order them by number.

The normal American breakfast is often thought of as a fruit juice, cereal, entree item, bread, and hot beverage, but these items are not always present. Continental breakfasts are common, which include a fruit or juice, coffee or tea, bread or rolls, and butter. Buffet service is often used for breakfast service with the buffet table holding fruits, juices, cereals, breads, etc. Bacon, ham, and sausages may be also offered and a cook may be able to prepare eggs to order.

There are a number of special occasion breakfasts. A wedding breakfast may begin with beverage alcohol such as a gin fizz or champagne, or a fruit punch. A fresh fruit cocktail is often the first course, followed by a breakfast entree and bread. If the breakfast is late in the day, the entree may be quite substantial. A beverage such as coffee is served. Wine service may also occur.

A hunt breakfast is an elaborate buffet that may include a number of meats and other dishes. The meal is bountiful and heavy. A brunch is a fairly heavy breakfast served late in the morning. It is supposed to be a combination of the foods served at lunch and breakfast. Some brunches can be quite elaborate. A wide number of fruits and juices, meats, egg dishes, breads, salads, desserts, and other items may be offered. It is not unusual to see champagne served. The amount of service given at these occasion breakfasts may vary. If it is a buffet service, the server might only clear dishes and serve the beverage.

Lunch

The foods served at lunch can vary considerably. Guests might have salad, or soup and beverage may be all that is desired. At other times a full lunch is wanted and this might be a first course, such as a cup of soup, followed by a main entree and dessert.

The type of establishment and the foods being offered will determine the table setting. This may be a full setting even though the guest may order only a limited amount of food. Other operations will not have a complete setting but may bring the desired eating utensils after the order is taken. American service is most common for lunch service, unless the dining it quite formal and the luncheon is occurring at a club or fine dining establishment. In such a case, the menu might be quite elaborate, but the tendency today is to simplify.

Group lunches are often a complete, but lighter meal consisting of a main dish, vegetables, salad, dessert, bread, and beverage. A first course, such as soup, may be served at a more formal meal. Often clubs will meet during the noon hour for lunch and hear a speaker. The service must be fast because the guests are often on a tight schedule. Normally, however, the entire luncheon may last one or two hours. Again the cover setting will be appropriate for the items served. When groups meet this way the service procedure is similar to that used at banquets.

Dinner

People generally have more time for dinner than for other meals and can use the meal as a period of relaxation after a day's work. Dinner usually includes a main entree, a starch item, a vegetable, and perhaps a salad. Bread and a beverage usually accompanies the meal. A dessert may or may not be wanted. People out to dine will often start the meal with an appetizer or soup.

People often dine out to celebrate an occasion and may want a bit more elaborate meal than they would normally consume. If this group eats in the regular dining room, the service is much like that given to other guests in the area. However, if the group meets in a special room, the service may take on some aspects of a banquet. It will all depend upon the meal and the service desired. Servers who can do something to make the meal a memorable one are always appreciated.

Tea

The service requirements for a tea will depend upon the type of tea and the number of people to be served. For small groups, the service can be somewhat like serving a family group at a meal, but for large groups, the service must take on the aspects of a banquet.

A tea may be just a quiet meeting of several or more friends who wish to get together and enjoy each other's company. In this case, service needs will be limited. The tea along with milk or cream, sugar or honey, and lemon slices will be brought, usually on a tray, and set on a small table. Cups, saucers, spoons, and small tea napkins should be provided on this table also. The server will pour the tea and bring it on a tray to each guest. The guest takes the tea and adds what is desired, taking also a napkin and spoon. If the group is quite small, the cups and other items along with the tray may be placed on a coffee table sitting between the guests and one will serve the tea and pass it. It is not unusual for a light cookie, tiny sandwich, or dessert plate to be included. It is important that servers see that guests are relieved of tea cups and other items as soon as they finish and also that the serving table is at all times kept filled with the necessary items.

There are low and high teas. The plainest low tea is simply tea with sugar or honey, milk or cream, and lemon. A more elaborate low tea might include cookies, mints, bonbons, or nuts, or may offer a dessert or fancy sandwiches. Low teas can be seated service or standing. Large group teas are almost always standing. A high tea is like the most elaborate low tea, but some heavier foods may be offered, so in a sense it becomes a light meal. (The British often have high tea late in the afternoon and then have a late dinner.) High teas are always seated service and are usually not practical for large groups.

Buffet Service

Regular Buffet Service

A buffet is a service concept that presents foods on a table. Guests progress along the length of the buffet serving themselves, or indicate what they desire to servers attending the tables. They vary considerably in the kinds of foods offered, when they are used, and in the table service required to serve guests.

A common buffet arrangement is one in which guests serve themselves vegetables and side items and then progress to the end of the buffet where a carver carves and serves the meat item or hot items and places them on the guest's plate. For a breakfast buffet, instead of carved meat the server may make waffles, hot cakes, or egg dishes. Following this method, the hot food is served last, immediately before the guest sits down to eat. Some buffets are arranged so only one kind of food is on one table. The main course may be on one table, the vegetables on another, desserts on another, etc. Guests go from table to table as they please. This prevents long lines. Such a service is called **shopping service**.

A banquet may offer a sit down dinner, then an elaborate buffet of different desserts, which is called a **sweet table**. In some buffet service, guests are provided a tray and a tray rail on which to slide their dishes, instead of carrying them in their hands. This latter methods is the one commonly used in cafeterias where guests get everything they need and bring it to the table. Tables may or may not be set with flatware. In some cases, again like cafeteria service, the silverware may be rolled in the napkin for pickup by the guests. The serving side of the table is often protected by a small clear glass panel over the foods. The glass is high enough for guests to reach under and get the food but it protects food from any germs spread by breathing or sneezing. Because of this, the panel is called a sneeze guard. Hot dishes are usually heated by electric lamps. Cold dishes may often be on a refrigerated table top. If only hot food is served on a plate, then the plate temperature should be warm, but not too hot to hold safely. Likewise, plates used for cold food should be chilled. If both hot and cold foods are served on the same plate, then it should be at room temperature.

Buffets have both advantages and disadvantages. Certain buffets can handle a large group of people quickly but the menu must be limited and the service time short. Thus, a hotel housing a large tour group that would be difficult to serve using traditional table service may be able to use buffet service for the group. Buffets also appeal to guests who are in a hurry, allowing them to select what they want and eat immediately. Buffets also require a smaller service staff than other service concepts. Unfortunately, buffets lead to a higher food waste since guests often take more than they can eat. In some cases, servings may be controlled by proportioning or other portion control methods.

It is extremely important that the foods on a buffet be attractively prepared and presented. Food of different heights can be arranged to give a variation or different items can be placed on a raised platform. The tasteful use of floral arrangements, mirrors, and fabric can add to the presentation. Colors should be bright, natural, clear, and varied. Garnishing can be somewhat elaborate, but not garish or overdone. Some operations use a buffet to feature an elaborate brunch on Sundays or holidays.

Three types of ethnic buffets are also used to attract trade or give a special emphasis to an event: a smorgasbord, a Russian buffet, and a French buffet.

The smorgasbord originated in the Scandinavian countries. It offers a full meal with many kinds of hot and cold foods, desserts, and beverages. To be a typical smorgasbord, pickled herring, rye bread, and mysost or ejetost cheese should be offered. A typical hot offering is Swedish meatballs. A Russian buffet is also an elaborate offering of hot and cold foods. There is often a fairly large offering of roasts and other cooked meats, including game. The buffet should offer rye bread with sweet butter and caviar in a glass or ice carving bowl. Vodka will also be served.

A French buffet is a fairly elaborate offering that begins with a course of hors d'oeuvres and appetizers. Patrons select what they want, and servers take it to their seats. The main buffet usually offers a fairly wide array of roasts and substantial dinner offerings. Again, patrons choose their food while the servers take the items selected to their seats. Wine is usually served as well as coffee or other beverages. A dessert table may be featured, or desserts may be arranged on voitures and brought to seated patrons for selection. French buffets are not common.

Recently the "**action station**" concept of buffet service has become increasingly popular. It consists of cooking or finishing some food items on medium sized pans over portable rechauds as the guests go through the buffet line. The foods items are served by the server from the pan or to the guests' plate Russian style. Most patrons are very receptive to this type of service as they feel their food is freshly cooked, as opposed to the common belief that buffet food may be allowed to sit for long periods of time. The action station concept can vary from a basic buffet line where a vegetable and chicken stir-fry or a pasta is offered, to a more elaborate setup where items such as shrimp scampi or beef tenderloin medallions are served to the guest standing in line. If the latter is adopted, servers must have a thorough knowledge of menu and preparation techniques. The equipment, tools, and ingredients needed are the same as in any tableside preparation methods typical of upscale restaurants. Obviously, this increases considerably the service level and the overall cost of the meal charge.

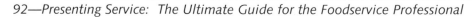

Cafeteria Service

Cafeteria service is a type of buffet service where the guest selects the items desired and then carries them away to be consumed elsewhere. Foods are arranged in the same sequence as a menu in a typical restaurant, therefore appetizers are usually first, then salads, pastas, vegetables, main items, and desserts. Servers behind counters serve the foods after the patron makes a selection. Many foods may be pre-portioned, such as appetizers, salads, and desserts, and these will be selected and picked up by patrons. Patrons pick up their own eating utensils and beverages using the self-service facilities.

Often the operation offers more than one counter in order to avoid long lines. A shopping center arrangement where different groups of foods are obtained from separate sections has been explained previously. In some cafeteria systems, different ethnic foods or regional specialties are obtained from different counter locations.

Payment for the foods selected is often made at the end of the selection counter at a cashier stand. Servers or buspersons will clear soiled items after guests leave, clean the table, and prepare it for the next patron. Occasionally they will also provide service on special requests.

Other Service

Drive-through Service

There are two kinds of service where food and beverages are ordered from a car: drive-through and car-hop. The latter is seldom seen today because of the increased labor cost over drive-through service, but it was the most common type used in the early days of take-out food.

Drive-through service is relatively simple. Customers drive to an ordering station where a menu is displayed. They use a two-way communication system to place their orders. They continue forward to the service window, and pay for and receive their order. The car drives away leaving the space for the next car. One can park and eat in their car and trash containers are usually handy for disposal needs. Often people on the road will stop and get food and eat as they are driving. This is particularly convenient for families with small children. An additional advantage of drive through is that during rush hour, when there is a long line inside the restaurant, service at the window is still reasonably fast. Many quick-service operations have extended their business hours and a great number of patrons prefer the drive through, particularly at very late hours, for reasons of safety.

Traditionally, management gives somewhat of a priority to drive through customers. Recently, some establishments have expanded the traditional system of two stops into three—one for ordering the food, one for payment, and one for pick up. Operators say the additional window allows for quicker service and helps maintain sanitation standards so that employees do not have to count money and handle food (or items that come in contact with food) at the same time.

Drive-through employees should be convivial and say good-bye while giving the food order to the customer. Using phrases such as: "Have a nice day," "Please drive carefully," and "Come back and see us again," will leave the customer with a pleasant feeling.

In this fast, mobile and vibrant society of ours, drive through food operations are increasingly popular. Some quick-service operations report that as much as 79 percent of total sales come from the drive through section of the business.

Room Service

Hotels and motels often offer room service, in which guests select foods and order them by phone. The food selections are then delivered to the guests' room.

Breakfasts can be preselected the night before by marking a menu left in the room and hanging it outside the door on the door knob. It is picked up during the night and the breakfast is delivered the next day at the desired time indicated on the card. Delivery carts for room service should be preset and ready to go so there is little delay. Some carts may have a unit on them that keeps hot foods hot and cold foods cold. Usually, however, covers are placed over the food to keep it at the desired temperature. Before leaving the kitchen, the server should check carefully to see that everything needed is on the cart, including all sauces, condiments, and utensils. If something is missing when the food is delivered at the room, the guest will not be able to eat immediately. Usually the guest places the food order by phone. The average time span between the ordering and the food delivery should be 15 to 30 minutes. Certain menu items might require more time to prepare. Often a special service elevator is used to deliver such foods.

Servers should check before knocking on the door. The room number is written on the guest check. After knocking, the server should identify himself or herself by saying "Room Service."

Once the guest opens the door, the proper approach is to address the guest and offer a sincere smile. When entering the room the server should always be preceded by the cart. In some operations the server has the option of setting the table in the room and placing the food items on it, leaving the tray with the food on it on the table, or leaving the cart in the room allowing the guest or guests to serve the food. Some room service carts are equipped with folding leaves. Once they are opened and locked into position, the cart becomes a full sized table. Because of the nature of this type of service, some guests are sensitive to privacy and expect the server to leave as soon as possible.

The server will ask the guest if there is any other request and will present the check. Over 90 percent of guests prefer the check to be included in the final room charges as opposed to paying cash. If the check does not indicate that gratuities have already been included, it us customary to tip the server directly according to standard tip rates.

Counter Service

Good counter service is not easy. It is fast work, and servers are often under considerable pressure. The turnover is short, multiplying the number of settings, orders, servings, and clearings. Guests are usually in a hurry and want fast service, and counter service is perfect for this purpose.

A counter station normally ranges from eight to twenty covers, and the work area for the server must be carefully planned so the service can go smoothly. Everything needed for settings and service should be within reach. At many counters, a place mat is used and the cover setting is placed on this with the knife and spoon to the guest's right and the forks and napkin to the left. Water is placed at the right above the tip of the knife. If other items are needed, they are placed at the cover after the order is taken. Condiments should also be close so all the server has to do is reach for them. Tote boxes or bus pans for soiled dishes may be close to the server's work area so the server does not have to go far to deposit soiled items. It is customary for buspersons to bring clean items and remove soiled ones. They may also help clean counters and maintain the area in other ways. Service can be speeded if the place where orders are placed and food is picked up are close to the counter area. If not, some way of getting orders into the kitchen without the server having to go there can be coordinated.

After a guest has been given the check—with a thank you—and has left, the soiled items should be quickly removed and a new setting put into place. A new guest may have already taken the seat expecting fast service. The clearing and cleaning of the counter top and placing of the new cover setting should be done with smooth, efficient motions.

In spite of the pressure of work, servers should remember to be cheerful, friendly, and considerate of guests' needs. At times when a counter server is not busy at the counter, side work should be done. If napkin holders are used, these should be filled, condiment tops should be wiped clean, salts and peppers set, and if need be, filled, along with sugar bowls. A failure to do good mise en place during slow times can make for a much more difficult and frustrating experience during busy ones.

Standup counter service is sometimes used to give fast service. Most often patrons select their food or beverage at one station and then carry the items to the counter, and bring what is needed. Often only fast, short-order foods are served in this style. A fountain or even a bar may be a part of such service.

Fountain Service

At one time ice cream parlors were very popular. They went into a decline several years ago, but now are beginning to regain popularity. Today the product comes usually in 2 1/2 gallon paper containers and prepared under good sanitary conditions. A much wider range of ice cream flavors and other frozen desserts are also available. The most common fountain items are sundaes, sodas, floats, milkshakes, splits, and carbonated drinks. The facility will usually have standard recipes outlining the proper way to make them and the correct ingredients and their amounts. Soft ice cream is very popular as well as frozen desserts made from yogurt or non-dairy products. In addition, there are sherbets, ices, frappes, milk shake mixes, etc. The frozen product also can be kept in units that produces and hold items at freezing temperatures. Spoons are the most common eating utensils used for fountain services and these are placed to the right of the cover. It is usual to see paper napkins used.

Chapter Summary

There are different styles of service, such as American, French, or English (butler). These service styles are explained more fully in the **Classic Service Styles Module**. These various styles are used in different service concepts such as buffets, counter, cafeteria, or banquet. While both are called a service, they are somewhat different in meaning. Sometimes service concepts are divided into occasion style of service, such as banquet, reception, tea, or breakfast, or into place style of service, where the service is distinctive to a place or location like a counter, showroom, or cafeteria.

Occasion services discussed in this chapter are banquet and reception. Place services discussed are buffet, cafeteria, counter, fountain, room, and showroom. Banquet service occurs usually at a dinner where a group gathers together to eat a common meal in celebration of something. The service style is usually American or Russian. Table service includes the kind of service given at breakfast, lunch, or dinner. Teas may be low or high; a low tea being simple with just tea and perhaps one or two simple food items; a high tea is much more formal and has a larger offering of foods that make it almost a meal. Receptions are almost always standup affairs. Service may take place from a buffet or servers may carry trays of beverages or small finger foods and offer them to guests. The trays on which such items are offered are called flying platters, having received this name from the Russians who invented this style of service.

Buffet service is perhaps the most common place service. It is a style of self-service with foods and beverages offered at a buffet, counter, or table. Some buffets, such as Russian, smorgasbord, or French, offer a distinctive type of food and beverage. Cafeteria service is another self-service distinctive to a place. Items are picked up at a counter—some are selected by customers and then served by servers behind the counter—and take to a table (or out of the operation) and consumed.

Counter service is a fast type of service where the guest is seated at a counter and orders the items desired, which are then brought and served at the counter. Good server organization is required because the service must be fast. Fountain service is a special type of counter service. Frozen desserts are featured, but beverages and other items may also be served. Room service is used in places where guests stay in their club, motel, or hotel rooms and order food. Servers must see that everything needed is on the delivery cart since the items are served far away from the place where they are prepared. Delivery time must be short so that hot foods remain hot and cold foods remain cold.

Showroom or cabaret service occurs when there is also entertainment. The timing of the service must be fast so as not to interfere with the entertainment program. In big showrooms or cabarets, the organization of the servers to see that there is fast, efficient, and courteous service is demanding. An aboyeur is often used in the kitchen.

Chapter Review

1. Define style of service, the occasion service concept, and the place service concept.

2. What is a banquet, and what styles of service are commonly used with it?

3. What are a smorgasbord, a Russian buffet, and a French buffet?

4. What is the difference between a high tea and a low tea?

5. What is a reception?

6. What is a flying platter?

7. What is a buffet?

8. What are some of the variations that can occur in cafeteria service?

9. What is room service, and what are some special challenges it poses?

10. How is the food order placed in drive-through service?

Service Areas and Equipment

Outline

Key Terms

Expediter

Réchaud

A la carte menu

Table d'hote menu

Du jour menu

Cycle menu

California menu

General menu

Learning Objectives

After reading this chapter, you should be able to:

- List and describe the equipment typically found in the dining area of an operation, and items used in table service.

- Describe the traditional hierarchy of a service staff.

Introduction

Servers use many kinds of service equipment and should know what these items are called and how they are used. Each cover, or a guest's place at the table, must have the proper equipment set in the proper way. Different service styles require different equipment.

Menus are a means of communication between guests and servers, and servers should know the items that are on them. Servers also work with two distinct staffs: the kitchen staff and the dining staff. They need to know the organization of each so they can work with them.

Dining Area Equipment

Personal Items

Servers should always carry an order pad or order checks, with a pen or pencil. Very formal service requires a serviette or service napkin. Matches or a lighter can be carried to light candles.

Service Stations

The service station, or wait station, assists the service staff in performing their duties. Depending on the type of the operation, a service station may be small with only a minimal amount of dishes and flatware, or it can take up a large portion of the floor space and have many types of equipment.

What is stored in a service station depends on your type of operation. Some items found in service stations include: flatware, dinnerware, table condiments, various paper goods, hot and cold prepared foods, glasses, cups, linens, trays, and tray stands.

All service stations perform one or more basic functions:

1. Storage for basic goods so the server does not have to make numerous trips to the main storage area for each customer's request.

2. A command and communication center where orders can be placed with the kitchen and management and the kitchen can communicate information to servers.

3. As a production area.

4. To hold items for cleaning the dining room.

In a banquet operation, the service station might be as simple as a table along a wall with extra silverware, napkins, bread and butter, and water and coffee pitchers. In a large chain quick-service operation, it might take up 50 percent of the floor space, including the area between the counter where the customers order, and the pass through area where the kitchen places prepared food before it is brought to the customer.

The service station, as a storage area, varies greatly from one operation to another. At this level its use is basic and the costs are minimal. However, larger operations have service areas that are more complex. They sometimes contain temperature storage units that hold both hot and cold prepared foods. The most common cold storage machines are ice bins, soda dispensers, and milk machines. They are found in most operations. Cold storage might also include dessert cases, or small reach-in refrigerators that hold premade salads, garnishes, or side dishes. Hot or warm storage are usually soup wells and bread warmers. In both hot and cold storage, the item is completed earlier in the kitchen, and the servers may have to plate or bowl the item before serving the customers.

SANITATION

The same sanitation procedures and time and temperature controls should be followed in the service station and in the kitchen. For example, it is illegal in most states to heat soup in a soup well. It should be heated to the proper temperature in the kitchen and then brought to the service station. The soup well should be preheated to the correct temperature beforehand.

As the costs of building restaurants increase, operators try to decrease the amount of space that does not directly have contact with the customer. Because of this, floor space in service stations is decreasing. To accommodate this change, restaurants are using vertical space—the space from the ceiling to the floor—to expand the usable space in service stations. It is important that managers are sure that proper sanitation and safety procedures are followed.

Command Center

In many foodservice operations, the service stations act as a command and communication center. POS systems allow servers to place orders to the kitchen in an orderly and traceable way. The POS system not only records and informs the kitchen of the customer's order, but prompts needed information such as appropriate temperature and side dishes, tracks open checks, counts customers, totals check sales, and monitors amount of sales. The service station is also where management and the kitchen communicate important information to the service staff. There might be a board that lists the soups of the day, specials, or what the kitchen might be out of. Management can use it to inform the service staff of schedules, section assignments, or special circumstances. In most operations, the service station allows the staff to communicate with each other.

Production Center

It is incorrect to assume that all production in a foodservice operation takes place in the kitchen. Many operations now require their service staff to be involved in some aspect of production. It might be as simple as making coffee and tea, or as complicated as preparing an entree-type salad. Beverages are most commonly filled in a service station. Almost all operations require the service staff to make coffee. Iced tea and lemonade are other simple beverages often produced by the service staff. Operations that offer malts and shakes usually require the servers to prepare them. With the growth of premium coffee and specialty drinks, the skills level and training required of the staff has increased. In operations that are food oriented but offer beverage alcohol, the servers might perform bartending duties. The necessary equipment must be located in the service station.

The production of food in a service area is usually limited to cold foods that are prepared beforehand, but do not hold well in their final state. These include ice cream, desserts with whipped cream, and salads. All the needed ingredients are prepared earlier by the kitchen staff or the server and assembled as ordered to ensure the highest quality product for the customer.

SANITATION

To ensure an item is produced properly in the service area or kitchen, procedures must be implemented and followed in training the staff on safety, sanitation, mise en place, and availability of equipment and materials needed.

Cleaning Area

The service station is sometimes used for storing cleaning items for the dining area. In some operations, the service station includes a place for dirty dishes/glasses, used linen, and garbage cans. Most also have sanitizing solutions to clean a table after it has been bussed. These solutions are used to clean a table after each use so it may be used, or turned, again.

Service stations or an attached area usually include the items needed for major cleaning after meal periods, at the end of the shift, or at closing. These include vacuums, polish, and cleaners.

SANITATION

It is important for sanitation and safety reasons to keep used dishes, glasses, linens, garbage, and cleaning and sanitation solutions away from clean items and food. Some states require handwashing sinks in service stations if food is being prepped there. It must be set up like a handwashing station in a kitchen and can only be used for handwashing.

Host or Hostess Station

All operations have an area where the customers first have contact with an employee. Most often this is a host/hostess station area. Most stations have a few basic items; phone, reservation book, menus, and a floor diagram (including open tables, server areas, and smoking/nonsmoking sections.) Some have POS systems that allow them to function like a cashier. Depending on the operation, it might combine some functions of a service station.

Kitchen or Service Area

The kitchen or service area is in direct contact with the kitchen. Usually a manager or an **expediter** is in charge of the area. The main function this person serves is to act as a communication link between the servers and kitchen staff (with the assistance of or instead of a POS system). If a server needs to communicate with the kitchen, they tell the expediter who relays the information to the kitchen. The expediter is also used if the kitchen needs to talk to a server about clarifying an order. Most operations that use an expediter require the service and kitchen staff to communicate through the expediter. They also help servers tray orders and assist in any production the server might have to perform.

Furniture

Tables

Tables for seated service are used in most operations. In others, counters with stools are used. Stand-up areas may only have high counters at which people stand while eating. Tables are usually sized to accommodate different sized groups, but may be put together for larger groups. Some square tables have hinged shelves which can be lifted up and locked into place to make a larger table, square or round. Table arrangements are dictated by serving needs and the dining area shape. Whatever the arrangement, it should be symmetrical and neat. Many tables are equipped with levelers; servers should know how to use these.

Chairs

Chair sizes and shapes will vary, but all should be very sturdy. Usually the chairs do not have arms. Chairs should be large enough so guests can sit comfortably. Chairs on rollers are easy to move, but may be dangerous because they can easily slip out from under a guest as they are about to sit or stand up. Many chairs are equipped with glides so they move easily over a surface. Servers should check these to see that they are secure. Also, they should check to see that the chairs are clean. Some high chairs may be needed for small children, while booster seats for larger children might be needed.

Other Furniture

A *guéridon* is a small mobile table used to hold a small heater called a *réchaud*, and food or liquid items. Serving dishes and eating utensils, as required, will also be on it. A réchaud is used in French service where much of the final preparation of food items occurs tableside. Various *voitures*, or small mobile carts, are used and can be rolled around the dining room carrying hors d'oeuvres, appetizers, salads, fruits and desserts, wines, liqueurs, and other items. Offerings should be neatly arranged. Some voitures may be refrigerated, while others may be heated and hold roasts and other items for carving and serving. A bus cart is used by servers or buspersons to hold used wares until they are transported to the warewashing area. Most operations do not allow the cart to be brought into the dining area when guests are present. Usually it is placed in an inconspicuous place and used wares are brought to it.

Tables are usually covered first with a silencer—a felt pad—which quiets the noises of dishes and utensils, and will absorb spilled liquids to protect the table. Cotton, linen, and ramie are the most commonly used fabrics for tablecloths.

Tablecloths should fit the table and hang down only about 8 inches. They should be plain and smoothly pressed.

Fine-dining operations may use naperones to cover the tablecloth. These are square, and do not cover the entire table.

Napkins, about 18 to 20 inches square, are often used in fine dining establishments for dinner. Smaller ones may be used for lunch or breakfast. In more casual dining, paper napkins may be used. Small cocktail or hors d'oeuvre paper napkins may be used at receptions, cocktail parties, or teas.

A serviette or hand napkin is usually made of cloth and hangs over the arm of the server to protect his hands when serving hot items and as a general serving towel. They should be used in all kinds of service situations. Place mats are individual covers, about 18 inches to 24 inches long and about 12 inches wide. They are made of various materials, and can be decorated with advertising or information the operation wants a guest to read, or games for children. Some place mats have the menu printed on them.

Table Service Equipment

Dishes

Dishware costs money and servers should give it good care. A dinner plate can cost $12 or more. Some operations prominently display the cost of individual dish items in the dishwashing area. A server who drops a tray of dishes costs the operation—or themselves—a lot of money. Sliding dishes and glassware from a tray onto a dishwasher table should never be allowed. Keeping flatware (forks, knives, spoons, etc.) separate from dishes is a must; some operations have a special sink or container where flatware can be dropped to soak before washing. Glassware should be given special care and kept separate from dishes.

Soiled ware may be loaded into dish baskets or tote boxes and taken to or from the kitchen. Servers who come to an area carrying a very heavy basket with no place to set it down can cause breakage or accidents. Know there is a landing space before you carry. Areas where soiled items are brought should have sufficient space so crowding and cross-traffic are avoided.

Most operations use a heavy reinforced china (porcelain) for dishware. Some may use lighter china, but this is often too fragile and expensive for ordinary use. It cannot stand the heavy wear and handling. Others may use ceramics or plastic and some paper. Whatever is used should be sanitary and clean. All china should be sparkling clean and sanitized when placed before guests. Glassware or china that is chipped, cracked, or has lost some of its glaze should not be used.

A description of some of the most used dishes follows.

Platter	Platters are long, oval dishes used to hold several portions of food; some smaller ones may be used instead of plates. Sizzling platters are almost always made of metal and are heated to a very high heat so the food sizzles on them as they are taken to the guests. Often an underliner is used. Guests should be warned not to touch them when the platter is set down.
Service/Show Plate	In fine dining operations, a plate larger than the dinner plate sits on the table at the cover when the guest is seated. Hors d'oeuvres, soups, and other first course foods in their serving dishes are placed on this plate. In many cases, management instructs servers to immediately remove the plate after the cocktail is served. Once removed, show plates are not used again. Show plates are usually 11 to 12 inches in diameter. They can be expensive and should be handled carefully. After their use, they should be wiped gently with a service napkin wet with vinegar and never brought to the dishwasher.
Dinner Plate	This is used for dinners and even lunch to hold the main part of the meal. It can also be used as a service plate for other purposes, and is usually 10 to 11 inches in diameter. In many establishments, oval shapes are used instead of the traditional round plate.
Salad Plate	A 7 to 8 inch diameter plate used to hold salads, soups in bowls, desserts, and other foods.

Bread Plate	Often called a bread and butter plate, it is usually 4 to 5 inches in diameter. Used for breads, rolls, toast, butter, jellies, jams, etc. Can be used as an underliner with a doily over it.
Demitasse	A round container with handle holding four-to-six ounces of hot beverage; a saucer is placed below the cup. Primarily used to serve espresso and specialty coffees.
Finger bowl	A cup-like container holding lukewarm water, usually with a slice of lemon or a rose petal floating in it. Used for lightly rinsing or washing the fingers. It is placed on an underliner with a doily on it. It may also be called a "monkey dish."
Cocotte	Small casserole for cooking and serving entree preparations. It often has a cover over it.
Cloche	A bell-shaped cover used with a serving dish; the cover holds in the heat. Can be made of glass, plastic, or metal.
Snail dish	Metal or ceramic stoneware with six deep indentations for holding snails cooked out of their shells. A shallower one with six indentations is used for snails in the shell.
Sauce boat	For serving gravies, au jus, and other liquids.

Glassware

All glassware should be sparkling clean and sanitized. **Exhibit 6.1** shows a water glass and a number of wine glasses. Often an all-purpose wine glass is used in place of these various wine glasses. **Exhibit 6.2** illustrates various brandy snifters and small liqueur glasses. **Exhibit 6.3** shows glasses used for other than beverage service. A carafe for holding wine or water is shown in **Exhibit 6.4**. A variety of beer glasses are shown in **Exhibit 6.5**.

Exhibit 6.1—Water and Wine Glasses

An all-purpose wine glass can be used instead.

Exhibit 6.2—Specialty Glasses

These liquor glasses are often used for serving specific kinds of beverage alcohol.

Exhibit 6.3—Glasses Used in Food Service

This parfait glass makes an ice-cream sundae look elegant.

Exhibit 6.4—Carafe

The carafe is useful for house wines.

Exhibit 6.5—Beer Glasses

Increase interest in your variety of beers by serving them in specialized glassware.

Service Utensils and Flatware

All foodservice operations have special utensils that are only used by service staff. They can be as simple as plastic tongs for portioning salads, or as formal as sterling silver serving sets. The most common utensils are ice scoops, tongs, ladles, stirrers, and spatulas. The utensils your servers need will depend on your operation. In any operation, the utensils must be in good shape, clean, and sanitized.

A list of some commonly used utensils follows.

Dinner fork	Used for main entrees and other foods eaten from a dinner plate. Can also be used as a general utility fork.
Salad fork	For salads, appetizers, some desserts, or fruits.
Fish fork	Used for fish and sometimes seafood dishes.
Cocktail fork	Used for seafood and other cocktails.
Lobster fork	For lobster when served in the shell.
Dessert fork	For pies, cakes, pastries, and other solid desserts.
Oyster fork	For eating clams, oysters, and other bivalves.
Fondue fork	A long fork used to pick up bread cubes and dip them into a cheese fondue; a shorter fork holds meat in hot oil for a meat fondue.
Teaspoon	Used for eating fruit sauces, puddings, fruits, etc.
Tablespoon	Larger than a teaspoon, used for soups, cereals, etc.
Soup spoon	Used mainly for soups.

Coffee spoon	For beverages, some cocktails, ice cream, etc.
Espresso spoon	For liquids served in a demitasse cup.
Sundae or iced tea spoon	For ice cream sundaes, ice sodas, tall iced beverages, etc.
Sauce spoon	A wide shallow spoon used for sauces and lifting foods out of casseroles, etc.
Snail tongs	For holding snails in the shell so the snail fork can extract them.
Lobster tongs	Used for holding lobsters in the shell.
Pastry tongs	For picking up and serving pastries.
Cake or pie server	For serving cakes, pies, pastries, tortes, etc.

Other Items

Some other items commonly used by servers are: pepper mills, water pitchers, coffee servers, teapots, bottle openers, oil and vinegar cruets, mustard holders, bread baskets, cheese graters, service trays, tray stands, check holders, and guest caddies.

Menus

Menus are important to service because they inform the guest what your operation offers. Most operations have menus in readable form. This may be a traditional individual menu, a sign listing specials, or a board above the counter. Some operations require servers to recite what the menu offers. This limits the menu size, but adds a certain atmosphere.

It is usual to print menus on hard paper called bristol or cover stock, but they can be on a sign board as used in a cafeteria or a drive-in. A place mat may hold the menu. Some operations have the menu on their order tickets.

Meal plans and menus differ. A meal plan lists the type of food served at each course of the meal, while the menu lists the exact items served. **Exhibit 6.6** shows this difference.

Exhibit 6.6—A Meal Plan and its Menu

MEAL PLAN	❧ MENU ❧
Appetizer	Melon au Kirsch
Soup	Lobster Bisque
Entree	Baked Ham, Champagne Sauce
Side Dishes	Sweet Potato, Roll, Harvard Beets
Salad	Pear and Grape with Honey French Dressing
Dessert	Hazelnut Cream

Menus may be designed for specific meals or occasions. Thus, we can have a breakfast, lunch, afternoon, dinner, or evening menu, a room service menu, or a banquet menu. Some may even be designed for the person, such as a children's menu, an early bird menu, or a vegetarian menu for guests with special dietary needs. Some menus are not intended to be used for ordering by guests, but only to inform cooks of preparation needs.

Menus are sometimes named by their nature. An **a la carte** menu prices all menu items individually. A **table d' hôte** is one that prices foods together in a group, often as a nearly complete meal. A **du jour** menu (du jour means of the day) lists foods only available that day and often there is little choice in selection—many times only one du jour meal is offered. A **cycle** menu is one that runs for a period of time with foods changing daily. Then the cycle is repeated. A **California** menu is one that lists snacks, breakfasts, lunches, and dinners all on one menu. A **general** menu is the main menu of a hospital from which special diets are planned. Thus, from the general menu the dietitian selects food which a specific patient might have.

It is customary to list foods on a menu in the order in which they are usually eaten. Others may even have a separate menu for appetizers or desserts. While less usual, menus may list different food groups by courses.

Menu items often carry some designator indicating how they are prepared. Thus, the menu may list a grilled cheese sandwich which means it is browned on both sides on a griddle. It is not toasted. However, the word grilled used with a steak usually means the steak is broiled over or under direct heat. A menu listing Prime Rib of Beef is not saying the meat is of prime grade but that the cut comes from the area known as the "prime" area of beef. Roasting means baking in dry heat while braising means cooking in a small amount of liquid. Boiled means cooking in a lot of water. Florentine means the item comes prepared in some way with spinach. Servers should know what terms of this type mean on the menu used for that day. If they do not know, they should look up the term or ask the chef or manager, so when guests ask they are prepared to give an accurate answer.

Servers likewise have to know the time it will take to prepare orders so they can inform guests on approximate waiting times and also plan their service time schedule for that particular table. **Exhibit 6.7** gives some preparation times, but these are averages, and may vary from the preparation time of a particular item on a menu of the facility in which one is serving.

Exhibit 6.7—Some Common Preparation Times (cooking times only)

BROILED STEAKS, LARGE SIRLOINS

Rare	20–24 minutes
Medium	28 minutes
Well done	28–32 minutes

LAMB CHOPS

Medium	12–25 minutes
Well done	14–30 minutes

PORK CHOPS

Well done	15–30 minutes

BARBECUED RIBS

Well done	45 minutes

CHICKEN

Broiled or sautéed	20–45 minutes

FISH AND SEAFOOD

Filets	6–10 minutes
Steaks	6–16 minutes
Whole	13–20 minutes
Baked oysters	12–16 minutes
Steamed clams	20 minutes
Boiled lobster	20 minutes
Broiled lobster, split	15–20 minutes
Broiled tails, split	8–10 minutes

The times in Exhibit 6.7 can vary according to cooking method. As servers gain experience, they begin to almost intuitively know about the time needed for menu items. However, if this experience is lacking, it is best to check.

Beverages, when offered on menus, are listed in various ways. Banquet or fine dining menus often list the wine to the right of the menu items with which it is served. Or the beverages served with the meal may be listed on the bottom of the menu:

Oysters Rockefeller

Boston Bean Potage, Croutons

Crab Mousse, Bercy Sauce

Grilled Loin Lamb Chops, Mint Sauce

Orange Glazed Carrots

Baked Stuffed Potato

Oriental Salad with Shiitake Mushrooms

Baked Stuffed Apple

❧🐜❧🐜❧🐜❧

Cocktails

Moet and Chandon Champagne

Cousino Masul Reservas Cabernet Sauvignon—1988

Chateau D' Y' chem Sauternes—1988

Liqueurs

Some operations may have a separate menu listing all beverage alcohol, usually aperitifs, cocktails, spirit drinks, brandies and liqueurs, after dinner drinks, beers, and nonalcohol drinks. The wine list usually separates domestic wines from foreign ones and in the order of listing in each is usually aperitif wines, dry white wines, dry red wines, sweet dessert wines, fortified wines, sparkling wines, and alcohol-free ones. Wines are often numbered and coded on the menu and guests can then order by the number. The server also lists on the check the wine by its number or code.

Service Staff

Two staffs are used to work in food preparation and serving: the production or kitchen staff, and the service staff. Servers are in frequent contact with the production staff where the type of staffing impacts what servers do in ordering and picking up orders. A thorough knowledge of kitchen staffing can help servers move more smoothly in the kitchen.

Kitchen Staff

Many food services operate with only one cook on a shift, along with a helper or two, and a dishwasher, and even the dishwasher can be missing. The cook usually directs the other kitchen workers, but frequently receives direction from an owner, manager, or assistant manager. If there is more than one cook, one is usually designated as head cook or chef, and this person is in charge, working under the direction of a manager. Kitchen staffs can grow to a considerable size with departments such as baking, salad, vegetable preparation, and cooking. Each department usually has a head that directs the work in that department; in many, the overseer of the kitchen may be a kitchen manager or chef. In some health facilities this may be a dietitian; dietary aids may assist the dietitian and also work closely with cooks in food preparation and in the dishing of foods and beverages to see that the foods meet the nutritional needs of the guests.

In the typical French organization there is often a continental kitchen. An executive chef is in charge, and is the overall manager of all food preparation and functions associated with it. A steward works with the chef in ordering, storage, and perhaps in menu planning. The steward is also in charge of much of the dining equipment, nappery, and other dining room equipment.

In food production, the sous chef runs the kitchen for the executive chef who usually has an office some place away from the kitchen. The continental kitchen is divided into various departments headed by chefs de parti, such as the *garde manger* (cold foods and pantry), *patisserie* (bake shop); *legumier* (cooked vegetable and garnishes), *potager* (soups and stews), *entremetier* (sauced and roasted dishes), etc. Under these chefs du parti will be assistants and helpers. Escoffier, the great chef, introduced this continental type of kitchen organization. He wanted to keep his kitchen quiet and so instead of having a number of servers calling orders to various unit, servers brought their orders to an *aboyeur* (announcer) who, after receiving an order from a server, called it out in a clear, loud voice so the cooks who had to prepare the items could hear.

Some kitchens use checkers to make sure orders are correct, to price orders, and to give them to the aboyeur to call out. It is the duty of the checker to check the food and beverages leaving the kitchen against the order to see that everything is correct, even the garnishes. In this manner, servers take from the kitchen only the correct order, nothing else. Some checkers act as cashiers, receiving money for the orders and giving back the proper change.

In addition other helpers, sanitarians, and employees will be on the staff. Servers do not come into contact with many of these employees. There are frequent opportunities for conflict when working under stress, and friendly, cordial relationships among employees can do so much to smooth over times when the work is highly demanding. Servers and kitchen staff work under the pressure of getting orders out promptly and in the standard required. A large number of orders may come in so fast that they pile up, and the kitchen staff may be under high pressure to get the food out in time and in proper quality. This is no time to make special demands. If not, tempers may flair. In some cases servers may share tips with cooks; this often creates a high degree of cooperation between cooks and servers.

Traditional Service Staff

Service staffs vary considerably. All servers are usually responsible to someone who heads the operation or a representative of management. Management hires, sets up schedules and stations, sets up service standards and training, and directs other service work. In larger organizations, some of these duties may be delegated.

Some operations such as drive-ins or quick-service operations have little or no staff organization. Orders are taken at a counter or window, and the food is brought there to give to guests.

Seated service operations are more complex. Usually a manager or some representative of management is in charge, with one or several servers working on a shift. If the service staff is large, a hostess or host may be in charge.

The most complex staff is modeled after that used in Europe, which grew out of the staffs used by Escoffier and Ritz. It is labor intensive and expensive, but elegant and lavish. The manager of serving is usually called *maitre d' hotel*, but may be called host or head waiter. Often this person is in charge of assigning stations and may even hire the serving staff. Under this leadership may be *captains* who are in charge of a group of servers, often called *chefs de rang*, who may also be called front servers. Each chef de rang is assigned a station numbering up to 25 seats. A *commis*, also called assistant waiter or back server, assists the chef de rang. Buspersons in this organization are often called *commis debarasseur*.

In some French server staffing the chef de service acts as the maitre d', and may have an assistant called the *chef d' étage* who directs serving in the dining areas. An individual called the *maitre d' hotel de care* is an individual who supervises a dining room section somewhat as captains do.

Other persons who may be included in the serving staff are the food and beverage manager, who is in charge of all food production and service. Banquet managers head up banquet catering staffs and are usually under the food and beverage manager. The wine steward or *sommelier* is responsible for wines and other beverage alcohols and their service. Some units use bartenders who only fill orders for servers, but usually do not serve the public.

In almost any type of service buspersons are used to bring trays of orders to a station, take away soiled items, and assist servers.

In a healthcare facility, a dietitian is usually in charge of service, but nurses who are not accountable to this person often deliver the food after it is sent from the kitchen. Servers may serve in visitor's or doctor's dining rooms.

Chapter Summary

Servers work with a lot of equipment and to serve properly must know what this equipment is and how it is to be used. In some operations the amount of knowledge required is much less than in others. A server working in a restaurant serving French food will use a wide variety of equipment, while one working in a drive-in will use a very different variety of equipment.

Each server needs certain personal items such as an order pad, pencil or pen, and other equipment needed to take orders and complete the serving of them.

Servers serve guests seated at tables or booths. These should be clean and neatly set. Service stations hold much equipment needed by servers. Servers should see they are properly stocked and are kept clean and orderly. In some food services mobile equipment may be used, and this also should be neat and clean, and not left where it can interfere with the work of serving. Other equipment will be used according to the serving needs and type of operation.

Tablecloths, napkins, naperones, place mats, and other napery may be required and servers need to know how to use these so that tables appear neat and well maintained.

Some servers have to handle a wide assortment of dishes, glassware, and eating utensils. The use of glassware or dishes with certain eating utensils is prescribed for certain foods or beverages and servers must know the proper ones to use. Different dishes, glassware, or eating utensils require specific placement at covers. Without a thorough knowledge of all the factors that go with the proper use of these items good service cannot occur.

Menus are used to communicate to guests the items available, and how they are prepared, along with the price. Servers need to know what the items are and how they are prepared so they can completely inform guests about menu offerings. Servers may have to explain menu items and also indicate preparation times. Servers have to be sales people as well as do the work of serving.

Servers need to know to whom they report and the role other employees on staff play in accomplishing food production and service. They need to know the kitchen organizations so they know how to place orders and pick them up. They also need to know this organization so they can work with the food preparation staffs.

Chapter Review

1. What is a service station? What is its use? What does it normally hold?

2. Match the terms on the left with their definition on the right.

   _____(1) Gueridon   a. Small casserole

   _____(2) Naperone   b. Used with a dish having small
                                              indentations in it

   _____(3) Snail tongs

   _____(4) Snifter   c. Covers tablecloth

   _____(5) Demi tasse   d. Holds about 4 ounces of espresso

   _____(6) Cocotte   e. Used to hold food and equipment for
                                              tableside preparation

                                             f. Holds a small amount of brandy

3. Match the terms on the left with their definition on the right.

   _____(1) Aboyeur   a. Assistant waiter

   _____(2) Maitre d' hotel   b. Front waiter

   _____(3) Commis debarasseur c. In charge of a group of waiters

   _____(4) Captain   d. Head of service

   _____(5) Commis   e. Busperson

   _____(6) Chef d' etage   f. In charge of wine service

   _____(7) Sommelier   g. Assistant to chef de service

   _____(8) Chef de rang   h. Calls out food orders

4. What is a cycle menu?

5. In what order are items usually presented on menus?

6. What kind of menus are not meant to be used by guests for ordering?

7. What is cover stock?

8. Explain the difference between a *guéridon* and a *voiture*.

9. What is the difference between an a la carte and a table d' hôte menu?

10. What advantages and disadvantages do you see in servers sharing tips with the cooks?

Serving the Meal

Outline

Key Terms

Tray jack

Food covers

Guest check

Pivot system

Guest-check system

Preset keyboard

Serviette

Guéridon

Crumber/crumb brush

Gratuities

POS computer

Greeter

Spindle method

Captain

Learning Objectives

After reading this chapter, you should be able to:

■   Oversee proper setting of tables, proper meal service, and clearing.

■   Describe receiving correct payment from customers based on accurate guest checks.

Introduction

Every profession has its rules and procedures for accomplishing required tasks; table service is no exception. Up to this point many general service rules and procedures have been given. This chapter focuses on the tasks of serving guests. First, we will cover some of the more general rules and techniques of handling trays and other service equipment. Next, rules and procedures for casual dining are covered, followed by special rules and procedures for formal dining.

Steps in Serving

Serving food and beverages involves a sequence of five steps: 1) greeting and seating guests, 2) taking the order, 3) serving, 4) clearing, and 5) presenting the check and saying good-bye. The tasks of each step vary according to different meals and type of operation. Thus, breakfast is served differently from dinner; a drive-in serves differently from a cafeteria; and counter service differs from table service. Service requirements also depend on whether guests want a leisurely or a quick meal. At breakfast, guests are usually in a hurry and things are done so that guests can be on their way. Lunches can be hurried or leisurely, and it is crucial for servers to take their cues from guests. At dinner, the pace is likely to be more leisurely. Banquet service, buffet service, and other specialized services have special requirements.

Greeting and Seating the Guests

The first employee that a customer comes in contact with represents the first opportunity to make your customer's experience a positive one. All employees, the owner, manager, host, and server, should know how to properly welcome a guest to the operation. How a customer is welcomed is dictated by the type of service your customer expects. Guests are typically greeted by a host or hostess or even the owner, but in other casual dining situations the greeter might be the server. The **greeter** should see that the greeting includes the most convivial elements: a pleasant attitude, a warm smile, eye contact, and a brief but welcoming phrase. If the guest has a reservation, it should be honored by immediate seating. However, this is not always possible and the guests might have to wait a few minutes. The procedure for handling this has been discussed previously under reservations.

Hosts and hostesses must be well trained in the operation's procedures for seating people. With regular clientele, their desires for a table may be known. Some like to be seated where they are seen. Others do not. Many restaurants have smoking and non-smoking sections, and guest preference should always be followed in this instance.

Hosts and hostesses should be alert and accommodating to guests' seating preferences. It is customary for hosts or servers to pull out the guest's chair so guests can seat themselves. Also be aware of special needs, such as those for people with disabilities and families with children.

The First Approach

The server's first contact with guests is crucial. At this point guests make a summary of what to expect in service and this often sets the size of the tip. The greeting, the seating, and the first approach to guests at the table should create a positive impression on the part of the guests and establish their estimate of the server's competence and ability to serve. The more pleasant one can make these first few minutes, the more likely it is that the server will have an easier and more pleasant time of serving, and that guests will enjoy their experience.

A friendly attitude is essential when dealing with guests, but the attitude should not be too familiar; some dignity should be observed. Do not indulge in unnecessary conversation or encourage familiarity.

The server or the **captain**, if there is one, should help guests with their coats or other items. In some operations, the server is not expected to seat guests, but in other fine-dining establishments, the captain and the server might do this.

Servers should see that their stations are ready to receive guests. Items such as candles, flowers, table tents, and place settings should be in place. The candle may or may not be lit, according to the operation. Some operations feel that upside-down glassware on the table indicates the table is not ready for guests, so it has glassware right side up when guests arrive. Flatware may or may not be on the table; in some operations, it might be placed only after orders are taken so the server knows what is required. In other operations, only a basic setting of flatware—knife, fork, and spoon—is set and the other items are added as needed. Unless an all-purpose wine glass is used, the proper wine glasses and other required glassware will be set after the wine order is taken. **Exhibit 7.1** shows various table settings. In full service establishments salt and pepper shakers may or may not be on the table. Condiments are not placed on the table until guests receive their food. However, coffee shops and other more casual dining units often have salt and pepper shakers and condiments on the table when guests are seated.

Exhibit 7.1—Table Settings

Servers should see that their tables are set appropriately and ready to receive guests.

Servers should be watchful of guests coming to their station. Let guests know you have seen them. Take a step forward to greet them and say the appropriate, "Good morning," "Good afternoon," or "Good evening" with a smile. If you know guests' names, use them. Take them to their table if the host or hostess does not do so, and ask if the table is suitable. At times this cannot be done because the server is busy with others. However, enough time should be taken to let the new guests know the server knows they are there. A short, "I'm sorry, we're so busy. I'll be with you in a minute." can help to give the server a chance to finish what is being done and come to the guests.

The Introduction

In some operations, when everyone is seated, the server introduces himself or herself by name, saying something like: " My name is _____. I will be your server this evening." Some operations do not like servers giving their names and only have the server greet the guest. If there are complementary snacks, they should be brought immediately, and water should be poured. It is also appropriate at this time to pre-bus or remove extra place settings from a table. If a table is set for six, but there are only four customers, the server should ask if additional guests are expected. If no other guests will be joining the party, remove the two extra settings. This gives the customers more room on the table, and saves the operation money.

Presenting Menus

Menus are frequently handed out right away, but some operations prefer to wait until the premeal beverage order is taken. If a premeal beverage menu is given out, the regular menu is then given out later. The person handing out menus depends on the operation. In most operations, the manager, or host, gives them out after the guests have been seated. Menus might already be placed at each cover. Or the server gives them out after the server's introduction or after the beverage order has been taken. The server should hand out menus to each guest's left, unless space does not permit this.

It is traditional to hand guests their menus, either opened or unopened. It is discourteous to drop menus on the table so guests have to pick them up and open them. An operation might or might not specify that menus be given first to women. The server also might hand out a wine list, or a sommelier will do this separately.

After the menus have been given out, the server should describe all food and beverage specials along with their prices. Servers should have tasted all specials so they describe them honestly and appetizingly. The server should also mention items on the menu that are especially good or unique, or that management wants to push. A remark such as, "The chef is particularly proud of this dish. He is planning on using the recipe for the next culinary contest," might arouse interest in an item. A description of the basic methods of preparation of some dishes also arouses interest.

Servers should be familiar with all menu items, their ingredients, and the methods used to prepare them. It is embarrassing, and bad for business, to not be able to answer guests' questions about the menu. Guests also may wish to receive information about the nutritional qualities of some dishes. However, servers should not give out information unless they are positive they know the correct answer. If a customer needs to know if the specific ingredients in a recipe for medical or dietary reasons, the server should ask the kitchen and give the customer the correct answer.

General Rules and Procedures for Serving

Proper serving is a craft that, when done correctly, flows so smoothly it appears simple to the untrained eye, yet when the novice attempts to do the tasks required, they become a challenge to one's knowledge and serving skills. What seems so simple when observed becomes extremely difficult when it has to be done. Just organizing the task alone so the service proceeds in a logical manner becomes a trial. However, with sincere application in learning the basics of service and by acquiring some dexterity in handling trays, china, glassware, and flatware, the novice server can become quite proficient in serving patrons, and in a short time can handle quite competently a fairly good sized station.

Servers cannot be considered professionals until they become well acquainted with handling all service equipment confidently. Although the restaurant industry is in constant evolution, the following discussion covers some of the recommended serving procedures and rules that apply to most forms of service. Proficiency in these procedures along with some others can go far in making a professional of the novice.

Some servers like to develop their own methods; some of these methods may be acceptable, but often they are not. The rules and procedures cited here have been tested over time and have been found to give the most satisfactory service.

As indicated throughout the various chapters in this book, there are many personal requirements of a server. The most basic one is the ability to get food and drinks back and forth between the kitchen and the guests. This is a demanding task.

It is essential that servers observe what is right and are not allowed to develop the wrong habits from the beginning. We are all creatures of habit and once we adopt certain methods and become accustomed to them, it is found that one learns to work smoothly, easily, efficiently, and quickly, giving a desirable level of service.

All dining operations should establish a flow pattern which servers should follow when moving in the dining area and kitchen. Breaking the flow can cause accidents because a fellow worker may not expect someone to act in a varying manner. Management should plan the flow pattern to follow, and train the service staff to follow it.

Serving Water and Ice

One of the first things done is to give guests a glass of water with ice. When pouring at the table, the pitcher should be positioned two to three inches away from the glass rim. (See **Exhibit 7.2**.) If it is too close there is a risk of touching and chipping the glass; if it is too far, even the most skillful server runs the risk of spilling. The glass should not be filled to the rim; two-thirds to three-fourths full is sufficient. Patrons dislike to handle a glass that is completely full. The busperson or server should be alert during the meal to see that water glasses are kept filled. Some operations provide a pitcher or carafe of water so guests may fill their own glasses.

 SPILLS

Make sure all employees know that spills require immediate attention. If the spill is in a busy area, an employee should remain there and direct traffic around the spill. The employee should warn nearby customers and fellow employees of the spill. While clean-up is in progress, the employee should post a sign, such as "Caution—Wet Floor." The sign must be left in place until the area is safe. If the spill is liquid and can not be cleaned up for a time, an approved absorbent compound may be used to contain the liquid. If water or chemicals are used in the clean-up process, the employee should avoid wetting any more area than necessary.

Exhibit 7.2—Pouring Water Properly

A water pitcher should be two to three inches above the glass rim.

Carrying Trays

Carrying a tray is one of the first tasks a server should learn. Trays come in various shapes and sizes, but the most common are the 27-inch-30-inch oval tray, used for large loads, and the 15-inch cocktail tray, used primarily to serve beverages. Trays are customarily carried by the left hand raised slightly above the left shoulder. (See **Exhibit 7.3**.) For sanitary reasons, the edge of the tray should be at least four inches away from the head or neck. The hair must not in any way come in contact with items.

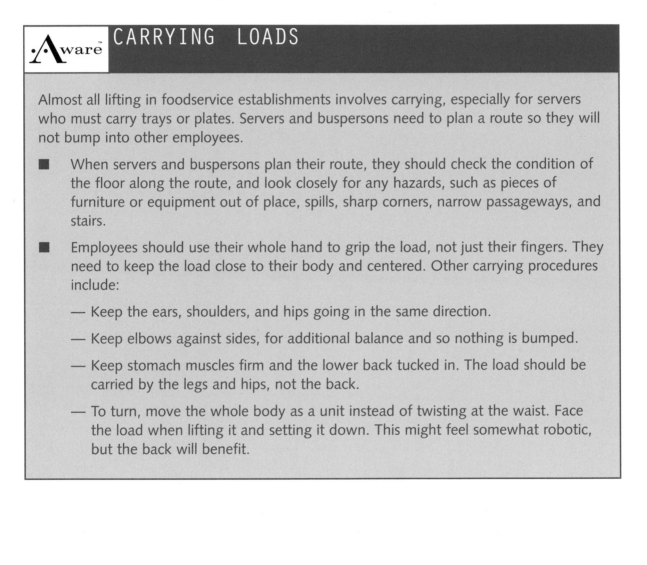

CARRYING LOADS

Almost all lifting in foodservice establishments involves carrying, especially for servers who must carry trays or plates. Servers and buspersons need to plan a route so they will not bump into other employees.

■ When servers and buspersons plan their route, they should check the condition of the floor along the route, and look closely for any hazards, such as pieces of furniture or equipment out of place, spills, sharp corners, narrow passageways, and stairs.

■ Employees should use their whole hand to grip the load, not just their fingers. They need to keep the load close to their body and centered. Other carrying procedures include:

— Keep the ears, shoulders, and hips going in the same direction.

— Keep elbows against sides, for additional balance and so nothing is bumped.

— Keep stomach muscles firm and the lower back tucked in. The load should be carried by the legs and hips, not the back.

— To turn, move the whole body as a unit instead of twisting at the waist. Face the load when lifting it and setting it down. This might feel somewhat robotic, but the back will benefit.

Exhibit 7.3—Carrying a Tray

A well-balanced tray is carried slightly above the shoulder.

It is not recommended to carry trays on the tips of the fingers. Using the palm of the hand gives more firm support and better control. (An exception might be when one is carrying a tray through a crowded area; raising the tray on the finger tips may help raise the tray high enough to get through.) The cocktail tray can be carried securely between the thumb and index finger, with added help from the other fingers, giving better control. This is especially true when carrying such a tray with glasses filled for a party of six or more. It is also recommended when carrying tall, fragile stemware or large items such as coffee pots.

Items on a smooth surface of a tray can slide; trays should be covered with cork, plastic, or a rubber mat to prevent this. A wet napkin spread over the surface of a smooth tray can also help reduce sliding. All trays should be kept clean for use.

Trays should not be placed on a table being cleared. Instead, the tray should be placed on a tray stand near the table being cleared. (A bus cart can replace the tray on the tray stand.) Trays should never be placed on a table at which guests are seated. They can be placed on a table, primarily in a large banquet or buffet operation, once guests have gone.

Some servers, especially in setting tables, like to carry stemware without trays by inserting the glasses by their stems through the fingers. While this is efficient and speeds table setting, for reasons of safety the use of a tray is recommended.

Perfect balance is the secret to carrying trays safely. Balance must be absolute to allow freedom of action and maneuverability when opening a door with the other hand, making turns, or just carrying the tray. Good balance can be obtained by distributing the weight of the items equally throughout the surface. The heaviest items should be placed in the tray's center with the lighter ones on the outside. When a tray is properly loaded by placing the palm of the hand exactly in the bottom center, there is no tilting from any side and the tray rests securely, with the fingers able to control any slight variation.

When delivering trays of food, it is recommended that a **tray jack** or stand be used. In some operations tray jacks are permanently set up ready to receive the loaded tray. In other establishments the server brings the jack to the table with the free hand, opens it next to the table, and places the tray upon it. If the tray is quite heavy, servers should place the jack prior to bringing out the tray. Sometimes another employee may come ahead of the server carrying the heavily loaded tray, open the jack, and help the employee place the tray on the stand. Once the tray is in place, service occurs with one or two dishes being removed from the tray at one time. When the tray is emptied, it is removed.

When resting a heavy tray on a tray jack, bend the knees slightly and put the tray down cautiously while holding it with both hands. (See **Exhibit 7.4**.) Many times it is better to ask a fellow server to help place the tray on the jack.

Exhibit 7.4—Setting Down a Tray

To set down a loaded tray, bend the knees slightly.

Loading trays requires attention, not only to achieve balance, but to get a maximum, but safe, load. This takes experience, and servers should note how more experienced servers load their trays. Food covers are used for many items. This allows another plate of similar size to be placed on top of the first plate. The cover also helps keep foods hot. It is possible using covers and stacking in this manner to get eight dishes on one tray. (In banquets, as many as 16 entrees may be put on one tray). Being able to stack in this manner gives good balance while at the same time helping the server to get more entrees to guests, saving travel time. It's also important to carefully load beverages so that the server can bring all the beverages to the guests in one trip, without spilling. (See **Exhibit 7.5**.)

Exhibit 7.5—Loading a Tray

Loading beverages carefully saves time.

Servers should take care not to place hot and cold items on top of one another or even allow them to touch. Pots holding hot liquids should not be filled so full that the hot liquid easily spills from the spouts when the tray is carried. Most experienced servers also do not put cups on saucers or dishes of varying sizes or shapes on top of one another. Such items fail to nest and can slide easily. When carrying filled containers that will be set on another saucer, dish, or tray, place the filled container on the tray and not the item on which it will sit. In this way, any spills occur on the tray and not on the item on which the container will sit.

Handling China and Glassware

Sometimes a server is seen carrying a tall stack of dishes with one hand underneath while holding the stack firmly against the chest with the other hand. This is both risky and unsanitary. Keep the fingers off the rim of the dish. If carried as a stack, put both hands underneath and hold the stack away from the body. A **serviette** or service towel may be wrapped around the stack to help hold it in place. Do not try to carry too high a stack.

Handle tumblers from their bases and stemware from the stems. Handle dishes with the hand under and with the thumb along the rim of the plate. Watch to see that soiled, chipped, or cracked china or glassware is never used. Fruit juice and cocktail glasses, cereal dishes, soup bowls, dessert dishes, and pots should be set on underliners when placed on the table. Set iced beverages also on an underliner or coaster. In some operations, underliners have a doily set on them before placing the item down. (See **Exhibit 7.6**.)

Exhibit 7.6—Carrying Glasses and Dishes

Never touch the part of a glass or dish that will come in contact with food or the guest's mouth.

If an item is hot while setting it before a guest, the server should use a fold-over-twice serviette or napkin between the thumb and index finger to protect the fingers. The server should also notify the guest that the item is very hot. If the guest does not hear the warning, the server should repeat the warning until the guest has acknowledged it.

Handling Flatware

Handling flatware properly is as important as handling china and glassware. Never touch the part of flatware that will go into a guest's mouth; hold only its handle. (See **Exhibit 7.7**.) All flatware should be wiped thoroughly before setting it on the table, not only during mise en place but also at other times. If flatware is extremely spotted, management should be notified so it can check dishwashing procedures. In particular, the blades of knives need to be wiped more than once; the spots from hard water and films left by cleaning agents and dishwashing machines are visible on the blades more than any other place. A cloth or towel wet with a bit of vinegar and water is good for cleaning; the scent of vinegar quickly disappears but the acid of the vinegar removes the alkaline agents that spot the object. Do not put bent, tarnished, or soiled flatware at a cover. Placing flatware properly on the table is an essential of good service.

Exhibit 7.7—Handling Flatware

Touch flatware only by the handle.

Some servers prefer to use a cocktail tray loaded with flatware organized so like ware is together when resetting tables. Others prefer to use flatware caddies loaded by kind in separate compartments. Whenever carrying flatware around the dining room, at least a plate with a napkin on it should be used. In handling spoons, forks, and knives, the fingers should touch only the lower part of the handle and never the business end.

Taking the Order

After the premeal beverages have been served and there are no reorders, the server should ask if the guests have made their selections. It is not unusual for some to have not even looked at the menu; the server should suggest appetizers and offer to return a little later. An alert server will note when the guests have seemingly made their choices, or when they may need assistance in making selections. The server should be there promptly to take care of guests' needs. Readiness to order is often indicated when a guest closes the menu and sets it down.

Guests may have difficulty in finding items or in reading the menu. Servers should note this and give assistance. Sometimes the guest cannot read well. The server may in this instance read the menu to the guest, explaining each item. Some may wish to know how items are prepared. Guests often want to know about a food's type, quality, grade, or preparation. For example, "Are the strawberries on the shortcake fresh or frozen?" "Are the oysters bluepoints or pacifics?" "Is the beef Choice or Prime?" "Are your pies baked here?" The time it takes to prepare might be important to some guests, and servers should be prepared to answer on any time questions. Servers must be prepared to answer a host of questions.

Some guests know immediately what they want; others may be undecided because they are not familiar with the service, are not hungry, have limited funds, or do not understand the menu. Servers can be a big help with these guests.

In most cases, the server takes the order, but in the more formal situations a captain might. Orders are often written on a **guest check**. These are usually serially numbered; servers are given a certain number of such checks, which they must sign for, before a shift. Some operations hold servers responsible for checks that disappear. There may be a standard charge levied against a server if a check is missing, which can occur with someone walking out without paying. Unused checks at shift's end are returned and credited to the server. However, where the POS system is used as it is in many operations, servers are no longer held responsible.

When taking the order, the server should ask whether the guests want separate checks or one check. The server should stand at the left of the guest ordering. For small groups, the server may stand in one place where each guest can be seen and heard. Do not hover over guests. Be sure to get complete information, such as the kind of vegetable, the doneness of the meat, etc. If an item takes long to prepare, tell the guest. It is desirable that servers also try to learn who is to pay the check so that presenting the check is clear.

Servers use systems on their order pads to help them remember who ordered what. The most common method is to establish the cover directly pointing toward a set point in the dining room as cover 1 or A, and to number the others clockwise. This standard system of guest order location is often called the **pivot system. Exhibit 7.8** shows how three guest orders may be taken at a table seating four. Cover No. 1 starts with the guest always sitting with his or her back to the kitchen. When the server returns with the orders, the server knows that this cover gets the New York steak cooked medium rare with a baked potato, house salad, and green beans. Cover No. 2 gets the chicken fricassee, mashed potatoes, peas, and tomato salad. No one is sitting at cover No. 3, but cover No. 4 gets the seafood salad and toast. Beverage and dessert orders will follow; the server will code these properly so service is easy and fast. All servers should follow the same system. In this way other servers can step in and work the table without asking a lot of questions.

Exhibit 7.8—Check System

The pivot system is a standard method used to take guest orders.

Today much order taking is computerized. Servers send orders through the computer in the dining room to cooks in specific kitchen stations. It is possible for servers using this device to code in cover locations so that the server knows which guest ordered what when the orders are brought to the table. Usually the time the order is placed, the time it is ready, and the time it is picked up are recorded so management has a check on how long the flow of order taking to service takes and where to spot responsibility for delays. The computer system will also print sales tickets. Hand held computer systems are increasing in popularity, but they still require an operational system, like a pivot system, for taking orders.

In any **guest-check system**, servers should write out orders in a legible and organized manner. If second and third copies are made, the server should press hard so each copy is readable. Good copies and legibility can help management in making check-duplicate reconciliation or any other control system used. For clearer writing, a booklet or menu under the pad or check can be used. When writing on a plain pad, it is recommended that a line be drawn after each course. This indicates to cooks when items are desired in the meal. (There might be a request where the guest asks for the salad to be served before the soup, or the salad to be served with the entrees, etc.)

Abbreviations are necessary to save time. (See **Exhibit 7.9.**) However, they should be standardized. Every server should know and use the abbreviations used in the operation. Following are some common abbreviations:

Extra rare (blue)	XR	Very well done	VWD
Rare	R	Chicken sautéed	Ch saut
Medium rare	MR	Chicken fried	Ch f
Medium	M	Steak	St
Medium well	MW	Roast Beef	Rst B
Well done	WD	Prime Rib	PR

Exhibit 7.9—Guest Check with These Abbreviations

Using standard abbreviations saves time when taking orders.

GUEST CHECK

TABLE NO.	NO. PERSONS	CHECK NO. 339540	SERVER NO.	
St, MR				
PR, R				
2 bkd pot, sc on side				
2 cr spin				
TAX				

It is helpful to underline extreme degrees of doneness, such as extra rare and very well done. This reinforces the fact that the guest wants the item that way. The server should also note any special requests. At the beginning of the shift, abbreviations for specials will be given by management to both servers and kitchen staff. Abbreviations help speed up order taking. If they later cause confusion and errors, then their purpose is lost and it would be far better not to use them.

Always repeat the order after the guest gives it. This prevents errors and misunderstandings. Be sure to get all the information needed such as type of salad dressing, doneness of meat, and any special requests. Sometimes a guest may just give the entree order, close the menu and hand it to the order taker. It is then necessary for the order taker to find out what the guest desires for the rest of the meal, such as an appetizer, the kind of soup or salad desired, beverage, and perhaps even the dessert, although dessert orders are usually taken after everyone has finished with the entree.

Suggestive Selling

Servers do more than take orders, serve, and clear dishes. They are important sales people, and a great part of the sales function is suggestive selling. More than a sales tool, suggestive selling helps guests make up their minds so they are pleased with what they have selected. It is an essential part of excellent service.

Every encounter with a guest is different. You will have to ask questions and watch for cues to find out what guests like and dislike, how much time they have, and how much they want to spend. Remember that guests can neither see nor taste menu items before ordering them. They depend on servers to help them make the right decision.

Appropriate attitude, dress, and confidence will encourage guests to take your suggestions. Let guests know that you're there to please them, and that they can trust you. It is important is not to oversell; don't be pushy. This brings resentment and distrust. But don't be afraid to suggest; the bigger the check, the bigger the tip.

When taking orders, servers should use open-ended rather than closed-ended questions. Open-ended questions allow further discussion of the subject while closed-ended questions stop it. By asking an open-ended question, the server has a chance to discover something a guest might like. An open-ended question might be, "What do you like for dessert?" Even if the answer is "Nothing. I don't eat dessert," the server has a chance to make a sale by suggesting other

options, like a low-calorie fresh fruit plate, or perhaps a cordial or other non-dessert item. A closed-ended question might be, "Do you want an order of sautéed fresh mushrooms to go with your steak?" In this instance, the only response is "yes" or "no."

Here are some tips for effective suggestive selling.

- Suggest beverages and appetizers to start a meal.

- Suggest premium liquors when guests order generic drinks.

- Suggest fresh fruit if the guest hesitates on the desserts.

- Suggest side orders with entrees.

- Suggest definite menu items; don't ask, "Will there be anything else?"

- Know the menu and suggest low-calorie items when appropriate.

- If the order will take some time to prepare, suggest an appetizer.

- If the guest orders an a la carte item and it is also on the dinner menu, suggest the complete dinner.

- Use appetizing words, such as *steaming, sweet, spicy, juicy, fresh, savory,* and *refreshing.*

- Sometimes certain items are also sold for takeout, such as pies and cakes, salad dressings, and some prepared foods. If a guest particularly likes an item, an alert server recommends that the guest purchase the item to take home (for a bigger check).

- Suggest desserts, desserts to split, and after-dinner drinks.

Placing and Picking Up Orders

When orders are sent by computer to the kitchen, servers do not have to go there to see that the right cooks get their orders. Servers are given a key, code number, or authorizing card to enter the computer system. After entering, the server usually has an identifying number or code. The table number, number of guests, and the check number is usually entered after this is done. The time may be automatically recorded. The guest check is now inserted into the computer and the orders are entered. Often there is a **preset keyboard** with almost all items on the menu listed. All the server has to do is touch one of these keys to order items. In some cases, the machine may ask for further information such as doneness of meat, or flavor of ice cream, and the server must then add this information. When all orders are placed, the server punches a key to print out the guest check.

With manual systems, servers must still place their orders in the kitchen. In some fine-dining operations, the order in the kitchen is first checked by a checker, who then places the order. In formal dining, an *aboyeur*, or expediter, may take the order and place it with the proper sections. In some kitchens, servers call out their orders to the proper preparation personnel, but many avoid this because it can cause confusion and noise in the kitchen when several or more servers are calling out their orders. Quieter methods are often used. In the **spindle method**, servers put the order on a spindle for cooks to remove. In some operations, orders are placed on a rotating wheel, which keeps them in order as placed. The cooks then rotate the wheel to arrive at orders in sequence. In larger kitchens, servers must rewrite parts of the order so these separate parts can be placed in the proper section for preparation. Thus, a cold plate order might be separately written and go to the cold food section, the roast beef order rewritten to go to the steam table section, and a broiled steak order rewritten for the broiler section.

Various methods are used to notify servers that orders are ready. It is possible to have pagers send a signal to a specific server. Lights can be set in a dining area and when this light is on, the server knows the kitchen is signaling that the order is ready. One novel way is to have a large clock which can light up an assigned number of a server.

Before taking an order from the kitchen, the server should check to see that everything ordered is there and is correct. (In some cases a checker does this.) The server should take something out that does not appear right. Dished up food should be neatly placed; garnishes should be right and attractive. Food spills on dish rims should be wiped away. Do this with a serviette, or towel, in the kitchen and not in front of guests. Servers should use a serviette or napkin to pick up hot dishes.

Servers should not pick up a course of a meal and bring it into the dining area until the previous course has been finished and cleared. However, in some faster service operations it is acceptable to clear one course while serving the next.

Servers need to organize their orders on pick up so everything is on hand when service at the table begins. One must be sure in the kitchen to pick up the correct items. Different menu items can look very similar, and if in doubt servers should ask the cook which meal they ordered. In some cases, cold foods should be brought to the service station before the hot ones, to be ready for service. When the hot items are brought from the kitchen, the cold and hot items are then served together. Pick up may require getting order items from various kitchen sections. For example, hot entrees may be waiting on a counter under a heat lamp, hot rolls waiting in a roll or bread warmer, and garnishes and salads may be in a refrigerator. Pick up hot foods last. Good organization simplifies the service task at the table. When the pick up is complete, a last minute check should be made to see that the food is the correct temperature, and has an overall pleasing appearance and superior quality. Some operations use expediters who do the picking up and act as the link between service and the kitchen.

Serving the Guests

Serving the order is the total of all efforts of everyone involved in the operation. Much thought and labor go into the production of the items servers put before guests; none of this should be lost during service. As noted in Chapter 4, everything must be ready and in its place. All mise en place must be done before bringing food to guests. All supplementary serviceware must be at the server station.

Where specific food items are placed depends on the table setting used. First courses, soup, and appetizers of single servings are placed directly in front of the guest. In placing the entree, see that the main food item is in front of the guest. Appetizers to be shared are placed in the center of the table with appetizer plates placed before guests. Entrees are also placed directly in front of guests. Other items, such as bread and salads, are placed to the right or left of the guest. Side salads are placed to the left, while breads are placed to the right.

In more formal meals, the placement of certain items are more or less prescribed. Beverage glassware should be on the right. This text indicates these, but there is so much variation today, that one might say that there is no consistent standard. Follow the standards of the operation, and try to make things convenient for guests. Managers in setting standards should see they provide the type of service guests want and that servers are able to meet.

Servers, whenever possible, should use the left hand to place and remove dishes when serving at the guest's left, and the right hand when working at the guest's right. This allows the server to have free arm movement and avoid colliding with guests' arms. Never reach in front of guests, or reach across one guest to serve another. Present dishes from which guests serve themselves on the guest's left, holding the dish so the guest can conveniently help himself or herself. Set serving flatware on the right side of the dish with handles turned toward the guest so it is easy to pick up the item for self-service. In most operations, salt and pepper shakers, sugar bowls, and condiments are placed in the center of the table. In booths and on tables set against a wall, these items are placed on the wall side. Bread trays and baskets are placed in the table's center. Cups and saucers go to the right of the guest, with the cup handle to the right.

Normally, entrees are served with the main item placed on the lower part of the plate closest to the guest, called the **six o'clock position** on the plate. The garnish and whatever side items are on the plate should be neatly arranged to make an attractive presentation.

Breakfast Service

Breakfast service must usually be fast. Fruits and juices should be served chilled. Milk is required with cereals and some operations also offer a finer sugar than regular, called berry sugar. Toast should be freshly made, and buttered or not, as the guest indicates. Hot breads should be hot and fresh. Hot cakes and waffles should be served as soon as possible; they lose quality rapidly as they cool. Eggs cooked to order must be correct, and served immediately. Hot beverages should be very hot.

Many operations today offer buffet breakfasts. This helps guests move quickly through the meal and allows them to take what they wish. Some guests may not wish as much food as is offered on a buffet, so most operations find that in addition to the buffet breakfast, they still must offer a menu and allow guests to select what they want.

Lunch Service

A lunch may consist of only one dish, such as a soup or salad, and a beverage. In a luncheon with courses, each course is placed directly in front of the guest. Vegetable dishes are placed above the entree to the right. Salad is placed on the left, with bread to the left of the salad. If a chilled beverage is served, place it to the right and a bit below the water glass. The handles of cups on the saucer should be turned to the right at the three o'clock position. At formal luncheons, servers crumb the table between courses. (Crumbing is removing crumbs and other items from the table surface.)

Formal Dinner Service

The first course in a full-course meal is placed directly in front of the guest. If a cocktail fork or other utensil is needed, place it on the right side of the plate or to the right of the plate. If flatware is already in place for the rest of the meal, this appetizer utensil is placed to the right of these. If guests are to serve themselves from a dish, place appetizer plates in front of the guest to the left. Sometimes a salad is served as a first course instead of an appetizer. The placement is the same and, if served from a dish, service again is from the left.

In some cases a finger bowl may be served after the first course, especially after finger foods such as cracked crab. The water in the bowl should be warm and have a lemon slice floating in it. A new napkin should be offered.

Soup is typically served as a second course. The soup bowl is placed on a serving plate, and put directly in front of the guest. The soup spoon should be to the right. Offer crackers. Crumb between courses as needed.

Entrees are placed directly in front of the guest. A vegetable dish, if used, is placed above and to the right. Side salads are placed to the left of the forks. The butter plate is to the right above the knife. Sometimes guests serve themselves from a platter. Put an empty, warm dinner plate directly in front of the guest with the platter and serving utensils above this.

At the more formal dinners, the salad is served after the entree and is set directly in front of the guest after the entree has been removed. Sometimes a salad bowl is offered; if so, the salad plate goes directly in front of the guest with service from the left.

In some operations, a fresh fruit sorbet is automatically offered before the entree. It is usually served in a tulip champagne glass over a doily and underliner with a teaspoon. A flower petal or lemon wheel can be used to decorate the glass.

Often dessert utensils are not placed with the original table setting but are brought in with the dessert. The dessert is set directly in front of the guest. If the dessert is triangular in shape, such as a piece of pie, place the point towards the guest. Sometimes there may be some doubt as to whether a guest would like a spoon, fork, or some other utensil. In this case, place both at the guest's place and let the guest decide. (See **Exhibit 7.10**.)

Exhibit 7.10—Typical Dessert Setting

A typical dessert setting.

Booth Service

Booth service is often difficult because of the need to serve guests at the far end of the table. Service to guests seated there should come first to avoid having the server reach across dishes served to those closer to the server. Serving those at the far end of the table usually requires the server to lean over and reach to get the items in their proper place. To serve those on the right, place the left hip against the table and with the left hand, reach out and set the items down. For the guests on the left, place the right hip against the table and reach out with the right hand. In serving a small booth or a dish used by all, it is acceptable for the server to stand flat against the booth and reach over. Because booths often have less space than tables, it may be necessary to clear items as soon as possible. (See **Exhibit 7.11**.)

Exhibit 7.11—Booth Service

Booth service is unique in that it forces servers to violate traditional service rules.

Clearing Tables

In some operations clearing is not done until all guests have finished the course. In other operations it is acceptable to clear as each guest finishes. Some guests do not like to have dirty dishes in front of them and may ask for their removal. In such a case, the request is followed in spite of house rules. Servers should watch to see if guests wish their dishes to be cleared. Some signal this by setting their eating utensils down on the plate. If in doubt, the server should ask the guest if he has finished.

Clearing is done to the right using the right hand. Left-handed servers can make an exception to this rule if they don't feel comfortable handling certain items, such as heavy entree plates.

If beverages are not finished, leave them at the table unless the guests want them to be removed. Offer after-dinner drinks. Leave hot tea and coffee on the table for the dessert course. Water glasses remain on the table as long as the guest remains; keep them filled. Cracker wrappers and other miscellaneous items should also be cleared.

A **crumber** or **crumb brush** is used to crumb a table, with or without a tablecloth. If these are not available, the blade of a dinner knife or a napkin folded into a roll shape will do the job. Do not sweep loose food particles onto the floor; use a plate or small tray covered with a napkin. The objective of crumbing is to get the table neat and clean and ready for dessert, coffee, and after-dinner drink service. (See **Exhibit 7.12**.)

Exhibit 7.12—Crumbing

The goal of crumbing is to clean the table before coffee, dessert, and after-dinner drinks are served.

Changing Table Linen

Clearing and setting table cloths properly is very important in making a good impression on guests. One recommended method goes as follows. Come to the table with a clean, folded cloth. Avoid holding it under the arm. Remove all items from the table onto a tray, cart, or service station counter. Place the clean cloth on top of the soiled one. Next, pull the soiled cloth toward the server or bus person doing the job so the hem of the soiled cloth is even with the far edge of the table. Unfold the fresh cloth by holding the center crease and the top hem between the index and middle finger. Then gently flip over the bottom section. Now hold the top edges of both cloths between the thumb and the index finger. By pulling on the opposite direction, the soiled cloth will slide and come out while the fresh cloth goes into place. Step back a few feet to make certain that the new cloth is centered. If necessary, adjust the hanging edges so they are centered. (See **Exhibit 7.13**.)

Exhibit 7.13—Laying a Tablecloth

Proper laying of table linen speeds up table setting.

This method allows servers and buspersons to change cloths without exposing the naked table surface (often unappealing to look at) to patrons.

Another popular technique is to first place tableware items on the far side of the table. Then, with the forefingers of either hand, hold the two corners of both the soiled and new cloth, pull up towards the center of the table, slightly raising both cloth's edges and lower the fingers down so as to fold the new cloth in half underneath the soiled one. Holding the corners of both cloths, fold them over again in the opposite direction. This should leave you with the soiled cloth folded in half underneath the clean one. Pull out the soiled cloth from underneath in one swift motion from the other side of the table. The tableware items are then placed back in the center.

There are other methods, some of which are faster and more suitable for a large-volume operation, but they involve uncovering a section of the table. A common one is to fold one edge of the soiled cloth and place on that corner the tableware items. Pull the cloth out, rest it temporarily on a chair, and insert the new cloth using the unfolding technique outlined above. Fold back the edge of the new cloth so as not to cover the tableware items. The cloth will now cover the table except for the corner where the tableware items are. Once these items are placed back in the center of the table, fold back the edge of the cloth so as to cover the table entirely. Adjust the hanging edges so that they are even on all sides. In this manner, the tableware items are never removed from the table, as opposed to the first two techniques listed above.

Presenting the Check and Saying Good-bye

When guests have finished and are ready to go, they resent having to wait for the check. The server, as well as the host or hostess, should be watching for signals from the guests that they are ready to leave, such as reaching for coats, purses, or packages. The last thing a server wants is a guest waving them down or shouting, "Please bring the check!" Don't let guests rise and start to

leave without the check, forcing you to rush over with the check. Difficulty getting the check can cause ill feelings, which might make guests decide not to come back and leave a paltry tip.

Present the check after the last clearing has occurred. It is an act of courtesy to ask, before presenting the check, if anything more is desired—even if the guest is seemingly in a hurry—so no oversight occurs.

The accurately totaled check may be presented in two ways. It may be brought to the table and placed face down on the table. This signals that the guest is to pay the cashier. It is appropriate to say something like, "Thank you. I have enjoyed serving you. Please come back again. You may pay the cashier."

In the second scenario, the server is expected to collect payment for the meal. The check is brought to the person paying the bill and is presented face down on a tray or plate. The guest leaves payment plus a tip and leaves. If the guests does not have the proper amount, an amount greater than the check total is left and the server deposits the right amount with the cashier and brings back the change on the tray or plate. The guest picks up the change and normally leaves a tip. Servers can encourage a good tip by bringing guests' change in some single dollar bills instead of only larger bills.

It is common to bring a receipt torn from the check for the guest. Servers should never manipulate the receipt amount to be larger than the actual bill as this is a form of fraud.

Payment today is often made by personal check or credit card. When payment is by personal check, the server usually must get approval for acceptance from a manager. Always read checks to see that they are completed accurately. Most operations ask for identification from the guest.

Credit cards are much more common. Most are readily accepted with no identification because payment is guaranteed by the bank or credit card company. Always check the card for a valid date, and get authorization from the credit card company. The guest must sign the credit card receipt; the name and signature must match those on the card. Tips are added to the receipt. The tip is then paid in cash to the server by management, which is reimbursed when the charge is paid.

It is unprofessional for a server to show in any way that a tip is expected. It is also unprofessional to show disappointment in the amount of the tip. Tips are given to show appreciation of good service. A tip lower than expected might be showing that the service was less than expected. Servers should know also that some guests are unfamiliar with American tip practices. In Europe, for example, tips (**gratuities**) are often included automatically in the check total. This is often true in the United States with larger groups.

GUEST CHECKS

Guest checks should be used by all servers and bartenders for every transaction. Guest checks should be ordered by the manager only, kept in a secure location, and issued in recorded lots. The supply of checks should be audited to ensure that none are missing. The essence of the guest check system is creating duplicate checks for each transaction.

Traditional cash register systems require the register operator to open the cash drawer to receive and dispense money. Enhanced point of sale systems reduce or eliminate this need, and may allow several order-entry terminals to feed orders to the kitchen, central register, and inventory system. When a POS computer is used, all payment is recorded automatically. If a payment is not recorded, the check remains active. Management can thus see from time to time which checks are still out and for how long. Walkouts can be discovered more quickly.

Many restaurants now use touch-screen **POS systems** that allow servers to close out their own checks and get them quickly to guests. These systems also process credit-card purchases through computer-system links with national credit-card companies.

Before guests leave, servers should be sure to thank them again and invite them to return. This lets them leave with a good feeling about the operation and an incentive to return.

Closing

When the serving period has ended, there are closing duties. Soiled linen needs to be counted, bagged, and placed in proper storage. All unused linen should be folded and stored with other clean linen. Unused clean flatware and other similar items should be returned to proper storage. Condiments should be wiped clean and stored properly. Salt and pepper shakers, sugar bowls, and other similar items may be placed on trays to take to a work station to be replenished. Butter should be covered and stored under refrigeration. The work station should be placed in order. Tables needing fresh linen should receive it and tables should be reset for the next meal, with clean glasses and flatware. (Some operations do not want overnight setting.)

There will be sidework that must be done. Usually management draws up a list of these tasks so servers are reminded about what needs to be done.

Closing after a serving period varies according to whether there is another meal period coming or it is the end of the day. It is important that those who close for the night see that those who come in to start the morning shift are not unduly hampered because the evening group did not properly prepare for the next meal.

Formal Dining

Formal dining is characterized by a number of factors, including:

- The dress code is usually more strict and formal.

- The food and service are usually quite elaborate.

- Menus are also quite elaborate.

- The decor is elegant.

- More servers are used.

- The atmosphere is reserved, quiet, and peaceful.

- The price will reflect the cost of these extras.

Generally speaking, the basic service mechanics mentioned for casual dining are also applicable in formal dining, but their execution is more elaborate. For example, butter pieces in casual dining are usually squares, while in formal dining it may be in the shapes of rosettes or flowery curls, requiring more delicate handling by servers. In casual dining, plates tend to be placed directly onto the table, while in formal dining dishes are placed by servers for each course.

When French service is used, much of the service is from the *guéridon*. American and Russian service are also used.

In formal service, wine is featured much more as a meal accompaniment; a sommelier is usually on staff. The service of wine is also more elaborate and formalized. All servers in fine dining must be well trained, and work might be specific to service positions. Thus, a chef de rang's tasks revolve mostly around the table, while the commis du rang will leave the table to bring food from the kitchen and do most of the serving. Captains may do special work such as deboning a fish or preparing Crêpes Suzette.

The Busperson's Role

After guests have been seated and menus have been given out, often the next person to make contact with guests is the busperson. Buspersons might be young, but some like the work and stay in it for a lifetime. The position can be professional and lead to many job satisfactions. Often, besides a minimum wage, the busperson receives 15 percent of the food server's gratuity. In other food services, gratuities are pooled and distributed equally between all service personnel.

The busperson's first responsibility is to provide water, butter, and perhaps bread after guests are seated and given menus. Bread baskets should be lined with a napkin because it is more sanitary, keeps the bread warm for a longer time, and looks appropriate and correct. Butter should be served cold but not on ice. (Once the butter softens and the ice cubes melt, it looks messy and unappetizing.) Most often, water is poured at the table, but water in glasses can be brought to the table on a cocktail tray. (This can be heavy when serving a party of six or more. The procedure for handling a heavy tray is discussed earlier in this chapter.) There are many other duties required of buspersons, including the following:

- Bring in foods from the kitchen.

- See that ice and water are always at hand.

- Assist food server in removing soiled items. Always separate china, glassware, and flatware. Soiled items should be stacked according to shape and size to allow more room and give proper support to stacks. (This allows more to be stacked on the tray and saves on trips. It also helps the dishwasher.)

- Provide any supplementary service items needed by guests.

- Assist servers in beverage service.

- Refill coffee and tea orders.

- Ensure all condiments and service supplies, such as cream, sugar, lemon, teaspoons, cups, and saucers are readily available.

- Help maintain linen, equipment, and serviceware in an orderly manner in the various storage areas.

- Be available at all times for guests' special requests.

- After guests' departure, reset tables.

Here are some guidelines for doing the busperson's job effectively:

- Keep all equipment organized (a place for everything and everything in its place).

- Avoid overstacking items to guard against breakage and accidents.

- Always keep safety in mind; act immediately when there is broken china, glass, or spilled liquids on the floor.

- Walk, don't run. Perform tasks in a systematic manner.

- Try not to travel empty-handed. There is usually something to be carried in or out of the dining room or to work areas.

- Do not engage in long conversations with guests unless encouraged by guests.

- Good communication skills help make the job flow more smoothly and easily. Work to have open communication with other serving personnel and kitchen workers.

- At the end of a shift, leave the stations immaculate. Wipe down all condiment containers, bread warmers, cutting boards, bus pans, and carts, and make certain that all soiled linen is counted and properly bagged. Act immediately when you notice that the restaurant is running low on cleaning and paper supplies, equipment, or serviceware. Some operations like to have tables set up and prepared for the next shift. Others do not. Whether this is to be done or not is a management decision.

Throughout the course of the meal, the busperson should constantly remember one of the most crucial components of service—anticipating guests' needs. Service cannot be of good quality if guest needs are not anticipated. Patrons can become exasperated if during the meal they have to continuously ask for more water, bread, and butter. The novice has to focus attention on the table until, after a little practice and experience, it will become second nature to spot these needs, even from a distance. The busperson's approach should be immediate and courteous.

Chapter Summary

Some general rules and procedures used in all types of operations include serving water and ice, carrying trays, handling china and glassware, and handling flatware.

Greeting and seating guests is the first step in good service. This task is very important since it gives a good impression to guests as to what is to follow. An important point in this first step in service is that servers should see that their station is ready to serve guests. This includes proper table setting, putting glasses upright ready for use, etc. The seating of guests should not be a random function; those who seat guests need to make quick and intuitive guesses as to where to place guests.

Servers should greet guests with a smile and a welcome that lets the guests know that the server is happy to serve them. It is proper for the server to introduce himself or herself and to hand out menus. Different operations present menus to guests differently. It is crucial that servers know the menu and are able to interpret it and describe items for guests. Often guests have questions or need help in making their selections, and the server is responsible for this. Servers should also be adept and comfortable with suggestive selling, not only to increase check totals and tips, but also to please guests.

Servers should use some system to write up orders so they know which guests placed which orders. Mechanized devices and computers transfer orders automatically to the kitchen and record sales. Order taking is made much easier and quicker when servers use abbreviations.

In noncomputerized operations, servers call out orders, place orders on a spindle or wheel, or give orders to an aboyeur or announcer.

The serving of the order is the culmination of all efforts of service and production. The main item of each course should be placed at the six o'clock position. Most items are served using the left hand on the guest's left, except beverages which are served using the right hand at the guest's right.

Some of the essential rules and procedures for serving breakfast and lunch are covered. In dinner service, there often is a specific sequence of courses. Courses will be placed by servers or guests might serve themselves from serving dishes. Special considerations are discussed for booth service. Items should be cleared properly from tables and the table crumbed. Several methods for laying a tablecloth are discussed.

When the guest pays a cashier, the server usually lays the check face down on the table near the person who is to pay it. Knowing who is to get the check is important since servers can cause some embarrassment if they do not know. If the server is to collect, the check is presented on a small tray or plate. The server receives payment and brings back any change along with a receipt. The guest then leaves the tip on the tray or plate. Credit card payment is common today. The server should always be sure to thank guests and invite them back.

Service does not end when guests leave. There is much to be done to be ready for the next service at that table. It must be reset, and the area made presentable. When closing out the meal, there are other things that must be done so as to be ready for the next meal to come. If the operation is to close for the night, other tasks must be done.

Formal service is characterized by how guests dress, the kind of place in which it occurs, special kinds of foods and menus, table preparation and service, and other factors. French service using the *guéridon* is common, but American and Russian service are seen. A few special rules and procedures for formal service are discussed.

The importance of the busperson in accomplishing service cannot be understated. Their tasks help make the meal flow smoothly. Servers often share tips with buspersons. The busperson's role can be summed up in the saying, " In every way, support the server."

Chapter Review

1. Where are heavy items placed when loading a tray? Where are lighter items placed?

2. What precaution must be taken when loading both hot and cold items onto a tray?

3. Why are the surfaces of trays often covered with cork? If a tray is not covered with cork, what can be done to give a similar effect?

4. Who normally first greets guests?

5. What is an appropriate way for a server to introduce herself or himself to guests at a table?

6. Why is it good for a server to help guests by suggestive selling? How can suggestive selling help the server?

7. If a guest orders a steak medium rare, how would you write it as an abbreviation?

8. Describe how a computer is used in placing orders. What else does a POS computer do?

9. How are items stacked properly on a tray or bus pan?

10. How should you present a check if the guest is to pay a cashier? If the guest were to pay the server?

Management's Role in Service

Outline

Key Terms

Standard

Line organization

Delegation

Participative leadership

Reverse pyramid

Service controls

Role-play

Learning Objectives

After reading this chapter, you should be able to:

- Describe management functions necessary to a successful operation.

- Explain how motivating and training servers helps an operation deliver excellent customer service.

Management's Responsibilities for Service

Good service comes not only from servers, but also from managers. Managers bear a great responsibility for establishing service standards, training servers, scheduling servers, and providing them with the equipment, tools, and environment they need to do their job well. The establishment of good service is a partnership task between managers and servers working together as a team to deliver it.

To have good service, management must set high service standards, communicate them to employees, see that they are met or exceeded, and support employees in their efforts through training and recognition. Unless this is done, servers are apt to establish their own standards and practice them. This may not give the best in service.

Establishing Service Standards

Standards can be defined as specific rules, principles, or measures established to guide employees in performing their duties consistently. With standards, management can measure and evaluate employee performance, and performance of the operation toward pleasing guests.

Service standards include service policies and service mechanics models such as how to pour water, how to deliver food to the table, how to set up a table, or prepare a service station. Management establishes service standards based on the type of operation and the quality of service management wants to achieve. In a fine dining environment the service standards for wine service are usually more extended and elaborate than the wine service standards are in an informal, family-style operation.

The first step in managing and delivering quality service is to set up and communicate service standards. Setting service standards involves seven distinct steps by mangers:

1. Set standards and describe them in detail.

2. Establish policies and procedures for accomplishing these standards. A **policy** is a plan or course of action to meet a standard. A **procedure** is the manner in which that plan is implemented.

3. Provide the necessary space, equipment, and environment to achieve standards.

4. Provide adequate training and guidance to servers and ensure that standards are met.

5. Review with employees periodically, so they know and understand how standards performance will be measured.

6. Train employees to perform specific tasks to meet standards. Follow up with checklists, sidework, tasks lists, and job descriptions.

7. Encourage and seek out employee feedback so management can be aware of problems that need correction.

Experienced managers are unanimous in reporting that the establishment of standards is a relatively simple task. The most difficult task is to train employees in those standards and then see they follow them consistently day after day. Often management establishes the standards and then delegates to secondary management the task of seeing that they are followed.

Standards vary, from the elegant service required in a fine-dining establishment to that appropriate for casual dining facilities. This book discusses typical full service standards, but every operation must interpret these to suit its guests. In other words, what is correct and what is not correct will depend upon the operation and situation. Specific standards have to be built for each operation, and it is management's responsibility to establish them.

Management Functions

Good service occurs when operations are well managed. A poorly run organization is unsettling to employees and leads to poor productivity and work performance. Seeing that an operation is properly run is a management responsibility and revolves around managers performing **five management functions**: planning, organizing, staffing, leading, and controlling.

Planning

The planning task starts with establishing a mission and goals for the organization and its people. This should be done with input from all employees, including servers. After all, who knows customers better than the people who serve them every day? All employees must have a stake in organizational goals or they won't be realized. Once goals are set, managers must see that realistic plans are developed, followed, and revised so these goals are met.

Organizing

Organizing applies to all of an organization's processes and resources, including its people. An organizational chart shows an operation's positions and their relationships to each other, including who reports to whom. (See **Exhibit 8.1**.) When responsibility and authority in an organization flow from the top down, it is know as a **line organization**. In practice, a hostess may be responsible for servers and to an assistant manager, who reports to a manager. A number of head waiters may be responsible to the hostess, and each head waiter may have a group of servers responsible to him or her. Large hotels tend to use a complex line organization, from food and beverage managers, to maitre d', to dining room captains, to chefs de rang (food servers), to commis (assistant servers), to bus persons. An important factor in line organization is that people should have to report to only one immediate superior. Reporting to more than one person and circumventing proper communication channels can cause needless frustration and miscommunication. For example, "unity of command", as defined by management theory, ensures that all employees throughout the operation follow the same policies and regulations.

If a number of employees have a problem with their supervisor, they should be encouraged to talk to another person with authority to effect results.

Many organizations are replacing line organization with an approach called "team effort". One version of this is the **reverse pyramid** in which the manager is at the bottom of the pyramid, supervisors are in the middle, and front-line employees are on the top. In this model, managers are seen as serving front-line employees so they can better serve customers. (See **Exhibit 8.2**.)

Exhibit 8.1—A Sample Organization Chart

An organizational chart displays an operation's positions and their relationships to each other.

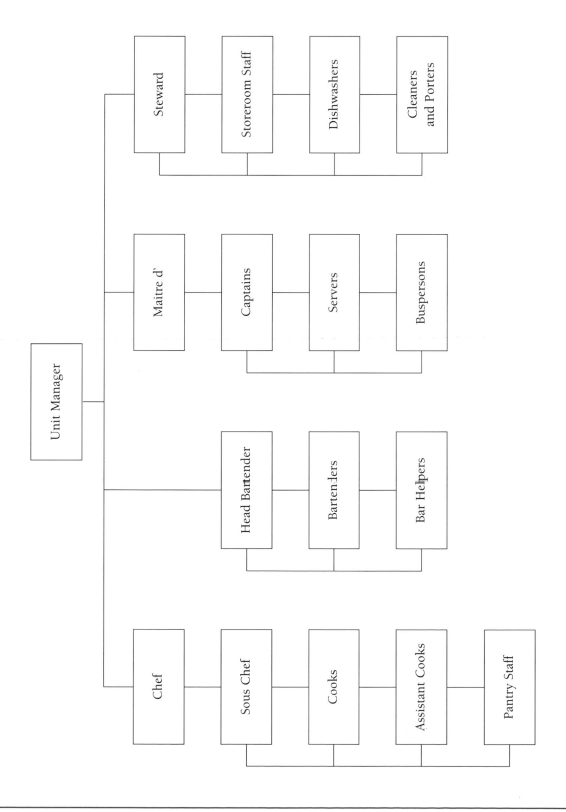

Exhibit 8.2—A Sample Reverse Pyramid Organization Chart

In the reverse pyramid, managers are at the bottom, with front-line employees and customers at the top.

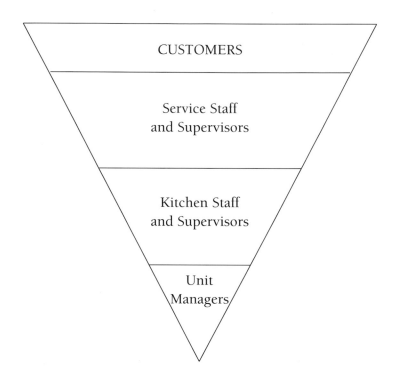

An important component of organization is **delegation**. Responsibilities and tasks can be delegated to employees, but the delegator does not escape responsibility for seeing that the job is done properly. An important part of delegating responsibility is also to give an employee authority to accomplish the work. Without authority, employees' hands are tied.

Proper delegation can work to the advantage of management in that:

- It can help the manager perform better on the job.

- It can be instrumental in establishing a definite sense of pride in the employee charged with the new responsibility.

- It can save managers time by freeing them from some tasks which will allow them to give that time to other important tasks.

Proper delegation requires that the person delegating is sure the delegatee:

- Knows how to do the delegated tasks.

- Is willing to do them.

- Is given the resources and all the assistance needed to do the job. One of these is that the delegatee be given the authority to do the tasks. If this is not given, usually the delegation tasks are not done or are improperly done.

Staffing

Hiring qualified employees is the first step in creating an excellent service staff. Hiring an employee that knows what is expected and will meet operation standards is more important than hiring an employee that simply acknowledges the specific duties of the job. Just because an employee has done the job before does not mean they will meet the hiring operation's standards. The person doing the hiring must ascertain whether those seeking positions have the necessary abilities and attitudes for the job, and reject those that do not have the necessary attributes.

One builds a staff by first determining what tasks must be performed to achieve the desired goal and then allocating positions so that these tasks are performed. After this, job specifications are written up indicating what the person in each position does. **Exhibit 8.3** is an example of a job description written up for a server. The next step is to hire people who can adequately fill these positions.

Exhibit 8.3 Job Description for a Server

Job descriptions indicate the necessary requirements for each position in the organization.

JOB TITLE	Server
JOB SUMMARY	Serves guests in all ways to insure maximum sales and profits.
	Coordinates service of meals to customers to insure 100 percent satisfaction, including troubleshooting and resolution of all complaints.
	Acts as primary bridge of communication between the operation's management and its customers.
JOB OBJECTIVES	Greets guests warmly and sincerely.
	Informs guests of specials, signature items, and specialties of the operation.
	Suggests menu items to guests and answers questions about the menu.
	Works with guests to solve all service-related problems.
	Keeps manager informed of potential and actual service problems.
	Gives guests' orders to the chef and other kitchen employees accurately and in a timely manner.
	Expedites orders and verifies that they are correct and complete before serving guests.
	Serves all menu items to customers in a timely and courteous manner according to the management's dictates.
	Provides continuous service to customers throughout the meal.
	Totals customer checks with 100 percent accuracy and presents checks to customers.
	Thanks all customers for their patronage and invites them back.

A staff consists of a group of individuals who are working together toward a common goal. A number of specialized positions are needed to perform the various tasks. In food service there will be a culinary staff, a server staff, a managing staff, and perhaps others. Within each staff will be various positions. For example, within the service staff the positions are a host or hostess, servers, bus persons and others. Each staff works together as a team to perform its functions, and the various staffs then blend their efforts together to achieve their goal. All servers should realize this and make themselves a part of the team. If they do, it will make their job a better one. Individuals working alone are without help. Being a member of a team helps to carry one along much more easily and smoothly.

Leading

An operation cannot succeed without effective leadership. To be a good leader one must know what the goal of the operation is and then take firm and adequate steps to reach that goal. Good leaders must command respect and goodwill, as well as success and profits. Good leadership motivates workers to do their best, and sees that workers are treated fairly and are given a chance to reach their goals.

Leadership styles run the gamut, from strict authoritarians to passive leaders. The recommended style today is a mixture of these, called **participative leadership**. In this style of leadership, the manager is a part of a team working with the employees toward a common goal. Traditionally, hospitality industry leaders have been authoritative, or dictatorial. There are reasons for this: 1) The nature of the industry was extremely competitive, with little time to promote a team atmosphere; 2) Managers had little time to train employees and had to set up tight control measures to see they were followed; 3) Because of the high employee turnover in the industry, managers were not able to build teams among their staff. As the labor market and the industry changed, managers found they had to change their management style to survive.

While authoritative leaders rule with little or no employee input and use fear of discharge or punishment to motivate action, participative leaders act as coaches to lead teams to success. Such leaders seek out ideas and opinions of workers and often follow them, giving employees a feeling of value and respect. They work alongside employees when necessary, creating a feeling of fellowship and the sharing of responsibilities.

Participative leaders still must ensure that standards are met and that rules are applied consistently and fairly. Participative leaders, rather than reprimanding and punishing, point out problems to employees and help correct them. This is why standards are so important, so that employees understand what is expected. It is important to specify the corrective action when pointing out mistakes, and to be as positive as is appropriate. Managers should correct the words, actions, or attitudes, not the employee personally. Counseling sessions should be held in private without any tones of anger or threats. They should be done more to make the employee feel they are being offered help rather than criticized.

Controlling

Controls lead an operation toward achieving its goals. Without controls, operations lose money and workers become frustrated and unhappy. Establishing standards and procedures for meeting them is key. Examples of **service controls** include:

- Greeting guests within one minute of being seated.

- Serving lunches in less than 30 minutes.

- Receiving fewer than 10 guest complaints per month.

- The specific steps to take when guests complain.

Servers and managers should check continuously to see that deviations from standards are avoided or corrected.

Motivating Servers

Management is responsible for seeing that employees are motivated and that they work in a motivating environment. To ensure good service, managers must develop a staff of servers that want to do the best possible job and do it consistently. Developing a motivated staff is not easy and management should make it one of its most important goals.

People are motivated to work by a number of factors. Primarily are the basic needs for food, housing, transportation, and so forth. Next comes the need to work in an environment that is safe and secure. Following this are social and environmental needs; people want to feel in any job situation that they belong and they are an accepted member of a group. A number of people also are motivated to work just to get a change in environment—elderly people may work because they are lonesome and bored with their home surroundings. Next comes the need to satisfy self or ego needs—people may like the independence of having their own income or just want the pride of having a job. And lastly, one may work to realize self-actualization or achievement needs—a teenager may become a bus person not only to have an income but to assert maturity.

It is a mistake to think that everyone is motivated in their work by all these needs. Some workers barely rise above basic security and safety needs, while others are motivated by all of them. Teenagers have different needs than more mature adults, and elderly workers often have different needs than those in other age groups.

Managers should not only try to satisfy the needs workers have, but also seek to extend them to higher levels. It will result in more satisfied employees. Every individual also has their own "pet" needs and, if management can satisfy these, a more satisfied worker is made.

Hiring Motivated Servers

As indicated previously, hiring is the first step to achieving motivated employees. Often the motivated employee can be detected in the interview by certain personality traits such as:

- A genuine interest and respect for people.
- A warm and outgoing attitude toward life.
- A sincere wish for the job.
- Habits which will not interfere with the job.
- A willingness to accept an entry-level position if they have little or no experience.
- A positive job record.
- Poise and confidence.
- An ability to solve problems and make good decisions.
- A pleasing appearance.
- Experience working in good operations, if applicable.
- Positive reasons for working other than income.
- Willingness to grow in the job and eventually be promotable.
- Good self-direction and self-reliance.

People must possess self-motivation to work hard and serve people enthusiastically. The following interview questions might help in evaluating an applicant's capability of becoming a well motivated server.

1. What was the biggest challenge you had in your current/last position? How did you meet that challenge?

2. What are your strengths? What are your weaknesses?

3. Have you ever had problems following instructions?

4. What type of goals have you set for yourself?

5. What do you see yourself doing five years from now?

6. Do you consider yourself a person that performs service duties with enthusiasm and dedication?

7. Do you like to deal with and be around people?

8. What do you expect from your co-workers?

9. What do you expect from your supervisors?

10. Can you point out how you introduced a new idea that improved your work or the work of others? How did it make you feel?

11. How would you handle a guest that is irate because the order was served late and the food was cold?

INTERVIEW QUESTIONS TO AVOID

1. Have you ever been arrested or accused of stealing?

2. Do you have a green card?

3. How long have you lived at this address? Where are your past residences?

4. Are you the wage earner/head of your household?

5. Are you married? What does your spouse do?

6. How many children do you have or plan to have?

7. Do you have a disability?

8. How old are you?

9. How many days of work did you miss last year?

10. What would you say your credit rating is?

Creating a Motivating Environment

One of the most important things managers must do to motivate employees is to see that a respectful, pleasant environment prevails. This helps servers feel both secure and appreciated, as well as motivated to give good service. One highly successful manager has said, "I treat my employees as I treat my customers; both are valued." A positive environment fosters team work and cooperation among servers. This is essential in the often pressure-filled, fast-paced, and demanding profession of serving.

To ensure good service, one must develop a staff of servers who want to do the best possible job, and do so consistently. One must create a feeling of respect and commitment between team

members, with managers participating in this team. Establishing standards and enhancing jobs such as empowering workers, or doing all one can to make the work easier are good motivators. Encouraging open communication, including servers in goal setting, treating employees fairly and with respect, and creating awards for good performance are others.

Some managers do things that destroy motivation such as:

- Assigning extra work without adequate compensation in praise, promotion, or other recognition.

- Unfair or nonuniform treatment or aggressive, abusive, cold, inconsistent, or stand-offish behavior.

- A failure to provide good communication, supplies, tools, or other standard conditions.

- Ridicule or sarcasm instead of constructive criticism.

- Oversupervision and a lack of independence and trust.

- A failure to follow up on problems mentioned by servers.

- An indecisiveness or unwillingness to support servers.

- Making decisions, without consulting servers, that may interfere with their being able to adequately do their jobs.

- Having too many bosses.

Empowering Employees

Empowerment is a management tool in which employees are given the power to make decisions that achieve higher standards of service. This power includes the authority and responsibility to make things happen. For example, instead of a server having to go to a superior to approve a meal deduction, the server is empowered to make the decision in order to correct the problem. Servers may be given some guidelines so they know what is and is not appropriate. Empowerment benefits servers, who are likely to feel more personal commitment to serving guests well. Empowered employees likewise benefit employers. Problems solved more quickly are better served. Dissatisfied customers are helped more quickly, and situations are resolved.

Rewards as Motivators

There are many ways management can let employees know when they are doing a good job and thus motivate them toward doing an even better one. Positive feedback and recognition of a job well done are very valuable. Monetary rewards and prizes may also be given. Many operations have an employee-of-the-month and employee-of-the-year award. Recognition of servers to guests is very powerful. Some operations offer servers bonuses after a certain length of employment. Others use extra vacation days, dining credits, concert tickets, amusement park passes, and trips to reward excellence performance. Achievement is traditionally rewarded with advancement.

Rewards should be open to all in the same category and all must be given complete information on what is required to get the reward. Monitor the program and evaluate it; does the program achieve its desired objective? Give the award as soon as possible after it has been won and tell them exactly why they qualify. Make the award worthwhile. The award program should be respected by the servers.

Things like offering adequate health care for servers and their dependents, giving educational assistance or assistance in personal problems, and even day care assistance can be good motivators and keep servers on the job. Whatever the incentive, managers should give them consistently and tie them to concrete behaviors. Special favors will undermine any incentive plan.

Remember that respect and loyalty cannot be purchased. It must be earned through sincere actions. Servers will see through attempts to be manipulated. Unless incentives are given appropriately, they will act as demotivators rather than motivators. Money alone rarely satisfies employees. In a survey taken of what employees want, the following were found to be mentioned frequently:

- ■ Job security
- ■ Good working conditions
- ■ Appreciation
- ■ Job satisfaction
- ■ Good wages
- ■ Acceptance
- ■ Dignity and respect
- ■ Good benefits

Note that only two (good wages, benefits) of the eight items are monetary; the other six deal with psychological factors that can cost little or nothing.

Unless servers get a sense of gratification from doing a good job, they lose interest. The work must have meaning to them and it must help them reach their own personal goals. Managers need to know that they cannot reach their goals unless employees reach theirs.

Evaluations as Motivators

The formal way of giving employees feedback on how well they are doing is the performance evaluation, or appraisal. Managers should make it clear that the purpose of evaluations are to help the employees develop and improve. It is important that servers know that evaluations are positive, and not negative. Evaluations should let employees know their opportunities for advancement.

The evaluator should look at customer comments, dollars produced, tips, covers served, and other factors to make the evaluation. Other intangible factors, such as attitude, teamwork, and punctuality should also be included. A performance record should be kept.

Appraisals can be helpful to management in giving information on employees attitudes, goals, needs and concerns. The information gained can also be helpful in giving advancement, pay increases, rewards, and what management must do to further the employee's pursuit of goals.

A good appraisal leaves one more sure about how well they are doing and what has to be done to improve. The things that need to be done for advancement, better pay, or other benefits can be learned.

Scheduling Servers

Proper scheduling can be a good motivating factor. It is essential that all workers be treated fairly and the same in scheduling. Some operations have a scheduling system that rotates servers among the different stations because some stations are easier to work or give better tips than others. Requests for special days off or certain hours should be given consideration and allowed, if possible.

Managers are responsible for seeing that adequate, competent servers are scheduled at the right times. If too little help is on the floor, good service is impossible. If underscheduling is consistently done, servers will slow down and management will create a very difficult problem for itself.

The number of covers in a station should be assigned according to the stations' distance from the kitchen. Team service allows at least one server per station to attend to guests while others do other work. Experienced bus persons also help in placing orders and delivering food and beverages.

Managers should take advantage of computer software programs that can help in planning better station assignments, scheduling, analyzing payroll, and monitoring service.

SAMPLE SERVER SCHEDULE

	Monday	Tuesday	Wednesday	Thursday	Friday	Saturday	Sunday
Jenn	Off	10 am–3 pm	Off	10 am–4 pm	10 am–3 pm	10 am–4:00	11 am–7 pm
Terry	Off	11 am–7 pm	10 am–4 pm	10 am–3 pm	Off	11 am–7 pm	10 am–3 pm
Damon	Off	10 am–4 pm	10 am–3 pm	Off	10 am–4 pm	10 am–3 pm	10 am–4 pm
Kate	Off	Off	4 pm–close	3 pm–close	3 pm–close	4 pm–close	3 pm–close
Carlos	Off	3 pm–close	11 am–7 pm	4 pm–close	11 am–7 pm	Off	4 pm–close
Lisa	Off	4 pm–close	3 pm–close	11 am–7 pm	4 pm–close	3 pm–close	Off
Deb	Off	3 pm–10 pm				↑	Off
Kyiel	Off	Off	4 pm–close	10 am–4 pm	4 pm–9 pm	4 pm–9 pm	4 pm–9 pm
Shawn	Off	11 am–7 pm	10 am–3 pm	6 am–11 am	3 pm–close	Off	5 pm–close
Rosa	Off	10 am–4 pm	Off	2 pm–9 pm	10 am–4 pm	11 am–6 pm	4 pm–10 pm
Dan	Off	4 pm–close	10 am–3 pm	Off	11 am–6 pm	5 pm–close	11 am–6 pm
Yi	Off	2 pm–8 pm	10 am–4 pm	4 pm–close	Off	5 pm–close	10 am–3 pm

Dining Room Arrangement

The dining area must be planned properly to allow for good service. Aisles through which servers must move while carrying heavy trays must be wide enough to allow for safe and good passage. Often managers are tempted to add too many tables, cutting down on space for servers. This is one good way to achieve poor service. Twelve square feet per guest is the minimum for regular table service and 20 square feet per guest is required for counter service. Club and luxury dining areas often have more than 12 square feet per guest. Banquet service requires less. Distances between tables should be 4 to 5 feet so that aisle space between servers is not restricted. The table should allow at least 24 inches of linear space per cover. Thus, a table for four should be at least 30 inches square.

Managers should see there is adequate room for tray stands, service stations, and other service equipment. One to four servers typically use one service station; the fewer the better. If possible, water should be piped to the station. In fine-dining operations, a carving station may be placed among tables for carving, deboning, and other activities. Mobile service or carving stations may be used. These can save space and give more flexibility in arranging tables.

Tray stands, service stations, and other things needed to give good service should be kept in order and be sufficient in number.

Kitchen Arrangement

Managers will assist servers by seeing that a smooth-flowing kitchen arrangement is set up. Unnecessary backtracking leads to delays, accidents, and frustration. A melee of workers going in all directions during busy periods makes things hectic for everyone, including cooks. A smooth, one-way flow, with pick-up arranged in sequence of courses, is best. This might be difficult to achieve since one section may be responsible for several courses. (A pantry might serve cold appetizers and salads while the grill area prepares only grilled items.) In these operations, sections should be placed apart from each other so as not to interfere with flow. Mobile equipment can be used to move items between stations, reducing server travel.

It is essential that food be delivered to guests with the proper appearance and temperature. Guests should not have to wait long for food. Efficient traffic patterns help ensure satisfied guests.

Training Servers

A wise manager would never send an untrained, inexperienced server out to work a station alone. Servers must be trained in what to do, how it is to be done, and the professional standard expected. While managers are responsible for seeing that servers are properly trained, the training function can be delegated to others who can do a good training job, who themselves must be taught to give good training.

While service training can be costly, poor service is more costly in terms of lost guests and profits. Here are some of the benefits of good training:

- Improves service and increases number of satisfied guests.

- Improves productivity, reducing both employee turnover and labor costs.

- Reduces waste, accidents, and breakage.

- Provides skilled, knowledgeable, confident servers opportunities for growth and advancement.

- Lowers frustrations.

- Reduces turnover.

Training should be an ongoing program in all operations because one never stops learning. Training should improve knowledge, skills and attitudes.

The Effective Trainer

A good trainer knows, people and how to teach them and motivate them to learn. Good learning can only take place when the trainer knows what must be taught and can teach this to the trainees. Before one starts to teach, one should have a lesson plan. This helps the trainer to stay on the objective and not stray, helps keep the program on time, and organizes the training session. Trainers need to be good leaders and must be able to control the learning session. They should be adept in building a feeling of trust and confidence with those they teach. A good trainer is a good listener as well as speaker. Students should be stimulated to ask questions and become involved in the training session. The learning given should have meaning to the student. Besides indicating the "what," "how," and "when" for jobs, the "why" should also be given. A student knowing "why" is more apt to make the learning permanent than just hearing the "what," "how," and "when." Trainers should not be afraid to repeat; repetition reinforces learning. They should also seek feedback on how well the student has learned.

Trainers should know the jobs they teach thoroughly and be able to communicate such knowledge to others. Not everyone can be a good teacher. One must have a personality for it and like to do it. Persons who are to train others should be trained for it. While teaching may come naturally, one can improve on that natural ability by knowing some of the techniques of teaching.

The trainer should plan detailed training and orientation sessions. Giving a pretest enables trainers to identify in what areas employees need the most training. Never assume that a server, even those with extensive experience, knows everything. It is a good idea to test servers after a training session, and then periodically following the session. This ensures application of training on the job.

The Training Session

Lecturing, group discussion, role-playing, show-and-tell, and on-the-job training (OJT) are some of the ways used to train servers.

Lecturing is good when the material to be covered is short and not too technical. It is good for imparting broad and overall information. Any lecture session should be short. Listeners quickly tire of just listening. It is best to use lecturing with the other techniques of teaching.

In planning the format and preparing the material for a learning session, the trainer will greatly benefit by keeping in mind that trainees remember:

15% of what they hear

30% of what they see

50% of what they hear and see

85% of what they practice

Group discussions provide motivation, interest, and subject retention for those being trained. They can be stimulating and very beneficial in getting learner interaction. The trainer must first identify the discussion objective, relate the topic to the learning objective, and manage the discussion. Trainers should try to get the discussion started and then stay out of it, only helping from time to time to direct the discussion.

Role-play allows servers to see and practice what is to be learned. It gives servers a chance to review and criticize, and helps to change or strengthen attitudes, skills, or knowledge. The active participation teaches one to work with others. The first step in role play is to give the trainee a situation relating to their position, and then allow the trainee to work his/her way through it. Role-playing is best used where active, physical effort is needed, coupled with the demonstration of some skill and knowledge. The proper way to perform may be explained by the trainer and then the student or students allowed to go perform the task. A critique should follow. Effective learning occurs if the others in the class join in the critique.

A **show-and-tell** method gives students a chance to hear and see how something is done, perform it, and then see where improvement is needed from the trainer and perhaps others in the class. Such feedback reinforces learning. Show-and-tell usually is used in classroom situations.

Usually the steps in show-and-tell are:

1. Tell the server how to do the task.

2. Show what is to be done, how to do it, and why.

3. Have the server repeat the task, verbally explaining what is done and why.

4. The trainer and/or the class review the performance.

Observation and on-the-job training (OJT) usually occurs in the workplace under real conditions. It can be quite successful if done correctly. Observation allows trainees to follow the trainer through the tasks so they can see how they are correctly performed, and then under a real situation go through the same tasks. The trainer observes and later provides feedback in private. OJT must be done carefully so as not to lower the quality of service to guests. Trainers should be competent to train and know the correct ways work is to be done. Too often the trainee learns, but learns the wrong way because the trainer did not know how to do it.

The Training Space, Tools, and Equipment

Some operations have wisely established their own training manuals. These usually cover such areas as greeting guests, giving menus, taking orders, and sequence of service. Mise en place (work done to get ready for guests' arrival) information should cover linen arrangement, table set-up, sanitation and safety practices, and stacking service stations. Menus and preparation that servers need to complete orders should be included. Selling techniques should be discussed and employees trained to use them. **Job descriptions** should indicate servers' tasks and responsibilities. They are good learning materials because they describe all the tasks done.

Externally produced training booklets, videos, software packages, CD-ROMs, and posters are available from professional associations and publishers. They are usually of high quality, having been prepared by authorities on various service topics.

Training models and formats vary according to the size and type of operation. Among the most common are: programmed learning by exercises, on site demonstrations, classroom activities, co-op programs, apprenticeships, and on-the-job-training.

Reservations

Management is responsible for establishing the reservation system and seeing that it operates as desired. Often employees operate the reservation system and sometimes the employee may be a server.

The reservation taker should show warmth and cordiality in taking the reservation. This person provides the first impression of the operation, and this is important. Follow the greeting with the name of the operation, perhaps adding, "How may I help you?" The conversation should be short and precise. Check the reservation book before confirming a reservation, noting the date, time and party number. At the end of taking any reservation, the taker should thank the caller.

Some reservation takers tell guests to please inform them if plans change. No-shows can create problems. It is advisable to ask for a number where one can reach the guest, especially in the case of a large party. This number may be called on the day of the reservation to be sure the party is still planning to come. Some operations take a credit card number and may make a charge for a no-show. Sometimes a request for a reservation is made too far in advance. A note can be made of this and the guest can call at a later date. The seating preference should be noted, especially smoking or non-smoking. Other things to note on the reservation book are occasion, special menu request, and any special needs.

It is important that reservations be arranged to get the best turnover possible. It is usual to stagger reservations to avoid having everyone come in at one time. This avoids overloading kitchen and service staffs. Some operations refuse to take reservations because of no-shows, and cancellations. They usually have enough walk-in business to make the operation successful, and a reservations system would be a burden rather than a help.

To establish a reservation system, first determine the number of tables and the number of seats available, and if tables will be moved to include different sized groups. From this, set up a reservation chart. (See **Exhibit 8.4**.). Usually a table is filled at dinner time for 1 1/2 hours for a table of one or two, but for larger groups the time is longer. Of, course, the meal has much to do with it. Breakfasts have a fast turnover, lunch has a little longer, and dinner has the longest stay. However, a breakfast group having a meeting or meeting for some other purpose may hold for two or more hours. An understanding about how long tables will be filled will give the reservation person an idea of the timing of reservations.

Exhibit 8.4—Sample Reservation Chart

A reservation chart details where different tables and sections are located in the dining room.

LUNCH RESERVATIONS FOR JOHN PURDUE ROOM

Name	Number in party	Phone	Your initials	Smoking/ no-smoking	Special requests
11:30					

Name	Number in party	Phone	Your initials	Smoking/ no-smoking	Special requests
11:45					

Name	Number in party	Phone	Your initials	Smoking/ no-smoking	Special requests
12:00					

continues

Name	Number in party	Phone	Your initials	Smoking/ no-smoking	Special requests
12:15					

Name	Number in party	Phone	Your initials	Smoking/ no-smoking	Special requests
12:30					

It is not unusual to overbook to about 10% of the dining room capacity. This is done to take care of no-shows, cancellations, and a slow walk-in night. This can, at times, lead to problems of having too many guests arrive with no tables to receive them. When this happens, the following is suggested:

1. Be consistent in handling situations.

2. Ask the party to please wait a few minutes, saying they will be seated as soon as possible. Telling a little white lie that the guests at the table are staying a little longer than expected is all right. If the wait is long, perhaps invite the group into the lounge for a before dinner beverage. Or, say you are sorry, but to make up for the delay you will serve the party a complimentary bottle of wine.

3. Give the time of wait if possible, but be sure the time is as accurate as you can make it. Do not say, "It will be just a moment," if you know it will be longer. Nothing causes more frustration than being given a time and then having it be far off. A slight difference is not important. If one cannot give an accurate time, tell the party, "Your table will be ready as soon as the party seated there leaves."

4. When approaching several groups waiting to be seated, pleasantly tell them their table is ready and then reassure the others waiting that they have not been forgotten.

5. A seat card prepared for the specific needs of the establishment can be a very helpful tool. Upon seating the guests, the host, hostess, or maitre d' will give the card to the server. The card contains all the information the server needs to know in order to follow up during the course of the meal: name of the host, table number, time seated, eventual time frame, special celebration, special requests, dietary needs, etc.

A typical seating card might read:

> Mr. and Mrs. H. Johnson–2
>
> Time in: 7:30 p.m.
>
> Table #8
>
> Server: David
>
> Notes: Anniversary cake, Mrs. J. allergic to seafood

There are many other things management must do to see that good service occurs. Often these duties do not directly influence service but do so in an indirect way.

A manager is responsible for seeing that good employee records are maintained, that tips are paid to employees and recorded, that vacation time is appropriately planned, and that employee activity programs, such as the establishment of a ball team, a dance, or other events are fully encouraged and supported. Management should work to make the whole operation one cohesive family. It not only leads to more motivated employees, but to the profit of the operation.

Chapter Summary

Managers are responsible for good service and should help their employees achieve it. One of the things that can help develop good service is a well run operation. A poorly run operation frustrates employees. To run well, a management must properly practice the five functions of management—planning, organizing, staffing, leading, and controlling.

It is management's job to see that servers are motivated to achieve high standards of service. A part of this comes in the hiring, but management has available many motivating rewards it can use, and many of these cost nothing. A warm environment, and fair and helpful treatment can motivate employees. Management should be a member of a team that wants to achieve. Participative management is usually the most successful.

Management is responsible for proper scheduling so servers know where to be when needed. This is management's job. Management is also responsible for seeing that the dining room and kitchen arrangements are set up to encourage good service.

Training is essential for good service, even with experienced servers. Training programs should be set up so servers know their work and possess the skills required. Training is management's responsibility and should be ongoing.

Complaints are often handled by servers, but sometimes management must step in and handle difficult ones. Knowing how to handle complaints is essential to prevent dissatisfied customers. Complaints can often be avoided by recognizing when the potential for one exists. For this reason, it is strongly recommended that managers and supervisors spend time with the service staff discussing how anticipating customer needs leads to better service.

Reservation programs are also the responsibility of management but at their establishment others may do the reservation taking, and sometimes these people are servers. The reservation taker should get complete information such as date, time, number in party, and other information desired by the operation. Some operations refuse to take reservations because of no shows, cancellations, and other problems relating to a reservation system. The reservation taker should show warmth in taking the reservation and thank the caller when the reservation taking is finished.

Chapter Review

1. Why are standards important to good service? Who should establish them?

2. Describe line organization. What are some advantages? What other type of organization is used today instead?

3. What are the characteristics of a participative leader?

4. What are some techniques for hiring motivated servers?

5. Name several faults in dining room arrangement that can cause problems for servers. Name some in kitchen arrangement.

6. What are some benefits of having a good server training program?

7. What information should one get from a guest in making a reservation?

8. What should the host or hostess do if two parties arrive at the same time, with reservations, and there is only one table available?

9. One server is so busy with a large party that she hasn't been able to approach a table that was seated five minutes ago to take the order. What can the other servers do to help?

10. There are a number of duties management must perform to ensure that good service occurs. Name some of these.

Glossary

A la carte menu	Menu on which items are priced individually
Aboyeur	Announcer who receives orders from servers and places them with the kitchen staff
Action station	Buffet service consisting of cooking or "finishing" some food items in medium-sized pans over portable réchauds as guests go through the buffet line
Auguste Escoffier	One of history's greatest chefs, who, along with César Ritz, operated fine hotels featuring fine dining and excellent service
Boulanger	Parisian who opened the first restorante on the Rue des Poulies
California menu	Listing of snacks and breakfast, lunch, and dinner items, all on one menu
Captain	Supervisor of the chef de rang and commis de rang who may do special work, such as deboning a fish or preparing Crêpes Suzette
Catherine de Medici	Member of one of Europe's richest and most powerful families who, as queen, started the growth of lavish and elegant French dining standards
César Ritz	Hotelier, who, along with Auguste Escoffier, operated fine hotels featuring fine dining and excellent service
Civil Rights Act of 1964	Law barring discrimination against employees because of race, color, religion, sex, or national origin
Control states	States that handle the sale and distribution of liquor
Crumber/crumb brush	Instrument used to removed crumbs from tables
Cycle menu	Menu that offers foods that change daily, with the cycle repeated after a period of time
Delegation	Empowering employees with responsibility and authority so that they can accomplish various tasks
Dram shop laws	Laws that hold servers responsible to third parties injured or killed by intoxicated patrons

Du jour menu	Menu that is planned and written on a daily basis
EEOC	Equal Employment Opportunity Commission
Empowerment	Management tool by which employees are given the power to make decisions that achieve higher standards of service
Expediter	Employee who acts as a communication link between servers and kitchen staff
Fair Labor Standards Act	Law protecting workers between the ages of 40 and 70 from discharge because of age
Family and Medical Leave Act	Law requiring employers with 50 or more employees to offer up to 12 weeks of unpaid leave in any 12-month period for reasons related to family and personal health
Flying platter	Tray of beverages or hors d'oeuvres carried by servers in flying service
Flying service	Reception service in which beverages and hors d'oeuvres are presented on trays
Food covers	Lids that keep foods hot and allow for plates of similar sizes to be stacked on top of one another
Functions of management	Planning, organizing, staffing, leading, and controlling to ensure that an operation is properly run
General menu	Main menu of a hospital from which special diets are planned
Gratuities	Tips
Greeter	Host/hostess, owner, or server who welcomes guests
Grimod de la Reyniere	Editor of the first gourmet magazine
Guéridon	Small mobile table used to hold a réchaud, food, and liquid items
Guest check	Forms on which guests' orders are written
Guest-check system	Any system in which servers write out guests' orders in a legible and organized manner
Guilds	Associations formed by artisans and skilled tradesmen to help regulate the production and sale of their goods
Immigration Reform and Control Act	Law making it illegal to hire aliens not authorized to work in the United States
Job description	List of servers' tasks and responsibilities

Line organization	Establishment in which responsibility and authority flow from the top down
Mise en place	Getting everything ready for the job to be done, and keeping things in good order as work is done
Participative leadership	Leadership style in which managers act as coaches to lead teams to success
Pivot system	Standard system of guest order location
POS (point-of-sale) system	Computer system that tracks guest orders, sales, guest counts, and other internal information
Preset keyboard	Machine-operated system that requires only that the server touch a key to order an item
Privacy Act of 1974	Law which forbids employers from asking non-job related questions which might be discriminatory
Réchaud	Small heater used in French service to heat foods
Restorante	Type of establishment first introduced by the Parisian, Boulanger, who claimed that the soups and breads he served were healthful and could restore people's energy
Reverse pyramid	Version of the team effort approach in which managers are at the bottom of the pyramid and front-line employees are on top
Role-play	Training technique that allows servers to see and practice what is to be learned
Sanctuary	Premises where guests are treated well by their hosts and every effort is made to see that guests come to no harm
Service controls	Tools or techniques that lead an operation to achieving its goals
Serviette	Small towel used by servers
Shopping service	Type of buffet service arranged so that only one kind of food is on each table
Silencer	Felt-padded or plush material placed under a tablecloth to quiet the noises of dishes and utensils and absorb spilled liquids
Sneeze guard	Clear panels on the serving side of a buffet table that protect food from germs
Spindle method	Order-placing method in which servers put orders on a spindle for cooks to remove

Standards	Specific rules, principles, or measures established to guide employees in performing their duties consistently
Taberna vinaria	Small eating and drinking establishment of the ancient Romans, from which we get the word "tavern"
Table d'hôte menu	Menu on which items are priced together in a group, often as a complete meal
Thermopolium	Stone counter with holes in which foods were kept warm, found in ancient Roman towns like Pompeii
Traffic sheet	Form on which servers sign for order checks
Tray jack	Stand on which heavy trays are placed
Union	Representative organization that acts for servers
Voiture	Small, mobile cart
Wave system	Banquet service system in which the entire room becomes one large station

Appendix A

What are the Criteria for Outstanding Performance?

The following is a list of typical duties and tasks of a server from *Server Skill Standards: National Performance Criteria in the Foodservice Industry.* © CHRIE (Council on Hotel, Restaurant, and Institutional Education), 1995. Used by permission.

Duty Area: Seat the Customer

Task 1: Verify Table Setup

Task 2: Seat Customer to Preference

Task 3: Coordinate Seating With Host/ess

Task 4: Confirm Table Setup

Task 5: Accommodate Customer's Special Physical Needs

Duty Area: Serve Customer at the Table

Task 1: Assess and Anticipate Customer's Needs

Task 2: Communicate Daily Specials and Changes

Task 3: Communicate Product Ingredients

Task 4: Suggest and Promote Products

Task 5: Take Customer Order

Task 6: Verify Customer Identification

Task 7: Recommend Wines Table-Side

Task 8: Assess Customer's Level of Intoxication

Task 9: Serve Beer and Drinks

Task 10: Monitor and Maintain Food Holding Temperatures

Task 11: Operate Microwave

Task 12: Prepare Food Arrangement

Task 13: Inspect Quality of Order/Service Area Before Serving

Task 14: Process Special Food Requests

Task 15: Deliver Food and Drinks to Customer

Task 16: Clean and Clear Customer Table

Task 17: Assess Customer Satisfaction

Duty Area: Maintain Stock Of Supplies

Task 1: Wash and Sanitize Glassware

Task 2: Stock Server Line and Stand

Task 3: Prepare Condiments for Shift

Task 4: Return Items and Equipment to Storage

Task 5: Operate Coffee Machine

Duty Area: Process Sales

Task 1: Present Check and Collect Payment

Task 2: Operate Point-Of-Sale Equipment

Task 3: Process Complimentary Offers, Discounts, and Voids

Duty Area: Support Other Staff

Task 1: Answer Phones

Task 2: Make Reservations

Task 3: Assemble Carry-Out Order

Task 4: Summarize And Disclose Tips Weekly

Task 5: Expedite Customer Order

Task 6: Maintain Cash Banks/Summarize Daily Transactions

Appendix B

What Does the Individual Have to Know, Be, and Do?

Information that an employee needs in order to perform successfully is called job knowledge. There are two types of job knowledge, and an employee should have both in order to successfully perform the tasks their position requires: General Knowledge and Property Specific Knowledge. General Knowledge, indicated with an asterisk (*), refers to the type of knowledge applicable across a wide range of properties. Property Specific Knowledge refers to the type of knowledge applicable to the individual operation. Included here is a summary of both types of knowledge needed for success in the server position, as well as a list that links each task with the knowledge needed to perform that task.

The following list indicates job knowledge most frequently used by successful servers from *Server Skill Standards: National Performance Criteria in the Foodservice Industry.* © CHRIE (Council on Hotel, Restaurant, and Institutional Education), 1995. Used by permission.

Payment/Checkout

- Where/how to accumulate money during shift
- How to separate tenders into common categories
- Location and function of checkout form
- Credit card vouchers, checks and room charges as money
- Procedure for making payment (e.g., through server or cashier)
- Location of check presenters
- Procedure to charge customer rooms
- Procedure for applying discounts
- Credit verification procedures
- Reasons for making correct presentation of check to customer*

Special Service Requests

- Limitations on accommodating special requests

- Where to obtain additional information to answer customer questions

- Procedure for assisting customer's with special needs without causing ill feelings (including sight impaired, hearing impaired and customers speaking languages other than English)

- What ingredients are available for use in extraordinary menu requests

- Employees who speak languages other than English

- Signals given that require additional service*

- How to determine customer's language*

- Reasons customer asks about ingredients (e.g., allergies, health reasons)*

Taking An Order

- Restaurant procedure for taking orders, ladies first

- Procedure for entering an order into the computer

- What to include in physical description of food

- Use of pivot point system for taking/delivering orders*

- Importance of giving customer correct information*

- Typical questions that assist in determining food likes/dislikes*

Wine And Alcohol

- Location of the reserve wine list

- Procedure for use of the wine steward

- Wines that compliment all selections on the menu

- Procedure for opening and pouring wine

- Property limit for tolerance of intoxication actions

- Procedure for handling intoxication situations

- Identification of your drinks, if leaving the bar area is required

- Location of service bar

- Restaurant/location/procedure for flags and signature drinks

- Correct garnish for standard drinks

- Questions relevant to determine individuals tastes in wine*

- Outward signs to be alert for involving intoxicated customer*

Kitchen

- Suitable microwaveable containers
- Oven operation
- Location of thermometers
- Correct temperatures for various food items*
- Sanitary method to check food temperatures*

Paperwork

- Procedure for filling out paperwork
- Procedure for closing out paperwork
- Importance of full disclosure and implications for falsifying records*

Customer Feedback

- Procedure for dealing with customer suggestions
- System for filing and using customer comments
- Construction, use and tabulation of customer satisfaction surveys*

Taste Panel

- Purpose of taste panels*
- Purpose for taste panel participation*

Dishwashing

- Automatic dishwashing system for glassware
- Procedure for preparing detergent sink for use for glassware
- Procedure for preparing sanitizing sink for use for glassware
- Procedure for transferring used dishes to dishroom

Storage

- Location of supplies (coffee, condiments, glassware, etc.)
- Procedure for dispensing condiments
- Location of various stock needed for service area
- Proper placement of stock for final food preparation (i.e., garnishes, etc.)
- Minimum stock level acceptable (before causing disruption)

Assignment/Work Schedule

- Location of side work assignment schedule
- Method to determine employee assignment on the schedule

Cleaning/Recycling

- What items can be retained for reuse and what items need to be thrown away*
- Procedure to get skirting cleaned
- Sanitation requirements
- Procedure for removing crumbs from table
- Location of bus pans

Point-Of-Sale Equipment

- Location of point-of-sale equipment operating manual
- Item codes
- Service charges
- Transaction processing
- Running special reports
- Location of log-out functions
- Computer generated screen/computer functions
- Closing out voids
- Location of delete Key
- Location and description of transaction cancellation key

Table Setup

■ Procedure for table setup

Carry-out

■ Procedure for preparing various drinks for carry-out

■ Location and procedure for preparing desserts, salads, etc. for carry-out

■ Procedure for packaging drinks for carry-out

Placing An Order

■ Restaurant system for electronic input of orders

■ Procedure for advance orders

■ Procedure for placing an order with the kitchen

■ Cost and description for daily specials

Serving An Order

■ Procedure for removing items between courses

■ Types of holding equipment

■ Who is available and qualified to assist in expediting the order

■ Use of finger bowls and wipes for finger foods

■ Procedure for serving (i.e., from left to right)

Presentation Of Food

■ Procedure for presenting specific menu items

■ Food item setup

Reservation And Seating

■ Reservation system

■ Seating flexibility

Special Events

■ Procedure to assemble and stage needed equipment/service items

■ Qualifications of employees for assignments to assist with event

Identification

■ Acceptable forms of customer identification

■ Procedure to validate and employee ID

Index

C